Edu

AF647548

REVITALIZING SCIENCE CURRICULUM
An STS Approach

About the Author

Mona Yadav is a Lecturer in Education in the Development of Women's Studies, National Council of Educational Research and Training (NCERT), New Delhi. She has done her M.Sc. (Chemistry), M.Ed., M.Phil (Education) and Ph.D. (Science Education) with high grade points from Jamia Millia Islamia, New Delhi. She has published many articles in English and Hindi in academic journals of repute. Her areas of interest are science education, girl's education and women's development. Her book *Effectiveness of Mastery Learning: An Indian Experience* is under publication.

REVITALIZING SCIENCE CURRICULUM
An STS Approach

Mona Yadav

CONCEPT PUBLISHING COMPANY, NEW DELHI-110059

Cataloging in Publication Data—DK
Courtesy: D.K. Agencies (P) Ltd. <docinfo@dkagencies.com>

Yadav, Mona.
Revitalizing science curriculum : an STS approach / Mona Yadav.
p. cm.
Originally presented as the author's thesis (Ph. D.)—Jamia Millia Islamia.
Study conducted in Haryana, India.
Includes bibliographical references (p.)
Includes index.
ISBN-13: 9788180693991
ISBN-10: 8180693996

1. Science—Study and teaching (Elementary)—India— Haryana. 2. Science—India—Haryana—Textbooks. 3. Rural schools—Curricula—India—Haryana. 4. Curriculum change— Social aspects—India—Haryana. 5. Science—Social aspects—India—Haryana. I. Title. II. Title: Revitalizing science curriculum : an science, technology and society approach.

DDC 372.350 440 954 558 22

All rights reserved. No part of this work may be reproduced, stored in a retrieval system or transmitted in any form or by any means, electronic, mechanical, photocopying, micro-filming, recording or otherwise, without the prior written permission of the copyright owner and the publisher.

ISBN -13: 978-81-8069-399-1
ISBN -10: 81-8069-399-6

First Published 2007

© **Mona Yadav**

Published and Printed by
Ashok Kumar Mittal
Concept Publishing Company
A/15-16, Commercial Block, Mohan Garden,
New Delhi-110059 (INDIA)

Phones : 25351460, 25351794
Fax : 091-11-25357103
Email : publishing@conceptpub.com

Preface

STS teaching is not about knowing the application of science in society, but about developing students' attitudes and about ways of looking at problems and deriving personally significant solutions for them. It is about understanding the way that science works in its social context.

The present study aims at 'renovating the science curriculum for upper primary level on the lines of STS themes with specific reference to the rural schools of Haryana'. It attempts to validate the STS themes in the Indian context for upper primary stage and evaluate the science textbooks for upper primary stage in terms of the validated STS framework in science education. Incorporations were suggested in a few lessons of the classes VI to VIII and two lessons each were modified. The two modified lessons of class VIII were taught in the real classroom situation. The book is divided into eight chapters.

In the first chapter a brief review of the status of science education abroad and in India, along with reasons for the need of STS education, is presented as STS education has come to play a significant and influential role in our everyday life.

The second chapter deals with the theoretical framework of STS education which includes the definitions incorporated in the term STS, inception of STS, conceptual framework, curricular framework and some projects undertaken by various researchers on STS. The third chapter deals with the review of literature related to various aspects of the study. The fourth chapter covers the plan and design of the present study, includes the assumptions and objectives of the study, the variables, sampling, description of tools, procedure, scoring, statistical treatment and delimitations.

The subsequent three chapters, i.e. chapters V, VI and VII, are devoted to the variablewise analysis of the results along with a detailed theoretical description about their nature and significance in science education. These chapters give a detailed qualitative as well as quantitative analysis of the objectives of the study. In the light of the findings, suggestions have been offered for further improvement of science education. The investigator also spells out the specific problems which need further explorations in the last chapter.

The investigator believes that this study will be of help in improving the teaching of science in schools. The study may also serve as a torch bearer to future researches. I commend this text to the science teachers for whom the study was conducted. I trust that they will see within the text, despite its many drawbacks for which I take full responsibility, some items which they may find useful and even some elements which may inspire them to some new acts of teaching or learning. It may help them to develop their science and technological education in ways that are both satisfying to themselves and stimulating to their students.

They will know, as I do, that teaching can never stay the same, nor mark time. It is bound to change because our concerns with science, with education and with the students who pass through our classes, continually demand fresh answers to fresh questions. Through the struggle to provide these, our own understanding of science and of people, continues to grow.

Mona Yadav

Acknowledgements

It is indeed a matter of pride for me to have completed my Ph.D. work under the guidance and supervision of Dr. Aejaz Masih. No formal words of thanks or gratitude can represent my indebtedness to him for his invaluable help and scholarly, insightful and critical guidance starting from the formulation of the problem to the final shaping of the research report. I received utmost benefit from his advice and suggestions at each and every stage in spite of his very busy schedule and being over-burdened with work.

I feel deeply grateful to my respected teacher Prof. Mohmad Miyan whose invaluable advice on the present study inspired me to work on it. It was entirely on the influence of his enchanting personality that I was drawn towards my present academic discipline.

I am also obliged and thankful to all the members of the department, particularly Dr. (Mrs.) G.M. Miyan and Dr. (Miss) Najma Amin for their moral support and encouragement.

I express my heartiest gratitude to Prof. J.S. Gill, Prof. Kalra and Dr. Abdul Maroof Khan who constantly gave their valuable advice, cooperation and consistently inspired me for speedy completion of this work.

Words are inadequate to express my gratitude to my friend Mr. A.K. Kaushik, who has moved heaven and earth for helping me in completing this work, sacrificing his professional and personal schedules. I express my heartiest thanks to him.

I wish to express my sincerest thanks to my friends Dr. Rita Sharma, Dr. Venkatesan, Mrs. Indu Kumar, Mohmad Zamir, Dr. Gauri Srivastava who sacrificed their professional and personal schedules and provided ungrudging and timely help.

Thanks are also due to the experts from various organizations who extended their suggestions to improve and develop the various tests used in the study.

I wish to express my thanks to D.E.Os., heads of the schools, teachers and students who extended all possible co-operation in the data collection for the study.

I am also thankful to the librarians and other staff members of Dr. Z.H. Library and Faculty of Education Library of Jamia Millia Islamia, NCERT Library and ICSSR Documentation Centre, Jawaharlal Nehru University Library, American Library, British Council Library, Delhi University Library and DIET, Daryaganj Library for their co-operation and help.

I will fail in my duty if I fail to thank Ms. Romi Nagpal, Mr. Suresh Sharma, Mr. Arun Kumar, Mr. Virender Yadav, Mr. Y.S. Tanwar and Mr. R.K. Chakravorty for the nice typing and for the computer packages used for this work.

Last but not the least, I cannot afford to forget and express my indebtedness and gratitude to my grandmother, mother, father, brother Vipin, sisters Poonam and Anju, brother-in-law Mr. K.S. Yadav and my sweet little niece Ishita and nephew Jayant. During the course of the study I had to unfortunately neglect some of my duties towards them. But they did not complain or show even the slightest grudge or regret, on the contrary they always remained as sources of inspiration and support.

Mona Yadav

Contents

List of Tables

List of Figures

List of Graphs

1

Significance of Science-Technology-Society (STS)

Introduction

In a society, education is an extremely important enterprise because the final product of this enterprise is a responsive and responsible citizen. The advancement of a well educated citizenry is measured by the quality of education its citizens enjoy, because the quality of education has become the signature by which a community can hope to realize the full potential of itself as well as its members.

Education should prepare children and young people to live in the world of the future and the world of tomorrow. It will help them to cope with certain profound problems and handle situations that have critical scientific and technological dimensions. Most of these profound problems are global in nature. To cope with these situations in future a greater competence in decision-making with concern for the whole world will be necessary and the responsibility for developing the needed competencies for making wise decisions again resides in the area of education. It helps decision-making not only for the future but even in the present democratic world, where every individual directly or indirectly participates in decision-making. Even the scientific and technological advancement depends upon the will of people, both their will as decision-making citizens to support it and their will as individuals to choose careers in science.

The inter-dependency of science, technology and society is increasing day by day. The increasing trend of inter-dependency between science, technology and society has created new problems and complexities and led to questions like, what would be the world of tomorrow? And how will the children of today face the future?

Being a product of society, education can and should play a role to make the interlinkages between the three, i.e. science, technology and society smooth as shown in Fig. 1.1 below.

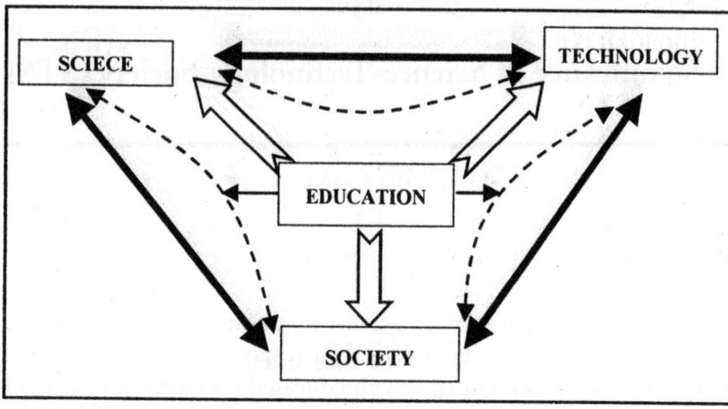

Fig. 1.1: Role of Education and Inter-dependency of Science, Technology and Society.

Source: Masih, Aejaz (1998). *New Trends in Science Curriculum*, New Delhi, Manak Publication, p. 3.

Looking back we see that the word, education, never had a common universal meaning. It was related to religious needs and had little or nothing to do with preparing children for active life. It was in the nineteenth century that education was generalized at primary level and made compulsory. Still education remained somewhat restricted in scope, focusing on the provision of reading and writing skills and the acquisition of religious knowledge. The schools created by colonial powers usually served a limited purpose, to provide trained staff of local people needed by the colonial administrators. In the twentieth century changes took place after the Second World War. Education was considered a right to be enjoyed by everyone and content was formulated. For the first time, in the twentieth century, it was felt that the quality of education depends upon the curriculum, which is the learning experience offered to pupils under the aegis of the school and yet goes beyond, the subject matter taught. In the later years of the twentieth century the curricula of lower schools was prepared to develop certain basic skills, content of all the subjects in the curriculum as reframed so as to be more useful in practical life.

Among all the subjects in the curriculum, science occupies a very eminent place at school and university stages of education. Science education, being an essential component of general education which is the need and right of every human being, plays an important role to make the new generation capable of handling problems in an effective way and enables them to take wise decisions in their future lives. It is required not only to satisfy personal needs or to fulfil national goals but also to adjust and cope with the problems of global nature which are going to intensify more in the days to come.

A brief review of the process of reforms in science education abroad and in India would be helpful to understand the present status of science education.

The Process of Reform in Science Education

The launching of the former Soviet Union's Sputnik in 1957 gave the impetus for curricular reforms in science education in the 1960s in the USA, UK and other western countries. The curricular reforms of the 1960s gave importance to how science is to be learnt, i.e. in terms of learning the process skills. These included observation, classification, experimentation, interpretation, prediction, hypothesizing, planning, investigation and communication. Another issue of the curriculum development programme was concerned with development of concepts as opposed to process skills. There was a continuous conflict between the process and content approach for the teaching of science. Neither of the two was independently sufficient, so the necessary inter-relation of concept learning and acquiring process skills was acknowledged as the new strategy. Hence, there is now an increasing interest in the development of approaches directed to achieving content and process goals in combination rather than separately (Young, 1988).

In these approaches, processes would be used to build concepts, while concepts would enable students to refine processes. Inspired by these developments and the changing needs of the children and of the society, a number of science improvement projects were started during the 1960s. Several innovative science education programmes like, Chemical Education Material Study (CHEM), Chemical Bond Approach (CBA), Biological Sciences Curriculum Study (BSCS), Nuffield Science Courses etc. came up in the United States, United

Kingdom and in some developing countries. These education programmes gave new orientation to science teaching. During the 1960s most schools began to develop or accept programmes that stressed on both the facets of science, conceptual aspects as well as methodological aspects.

The efforts continued further. The secondary science curriculum changed all over the world in the 1970s, and included a debate on whether science should be taught with an integrated approach or as specific subject. The main focus of the various programmes of curriculum reforms was on producing curricular material. Subsequently, teaching methods and teacher preparation continued to be given importance. Reforms in science education involved only the academic aspects of science. At the same time two other movements were gathering momentum. The first was the environmental movement and the second was the science, technology and society (STS) movement which acknowledged that science did not exist in a vacuum but that science, technology and society were inter-related. It was hoped that this would create informed citizens who would use the products of science and technology in a more responsible manner. A number of events in the world probably triggered these movements. The disillusionment caused by the tragedy of the Vietnam War compelled rethinking of all national curriculum development and focused on the negative aspects of science and technology (Yager, 1984).

Though many of the basic ideas and goals of the 1960s have been adopted by STS reformers, one reason for the disappointing realization of non-curriculum potential was the centralist approach to curriculum development that dominated the field in the 1960s. The teacher was to use the prescribed materials and often the curriculum projects aimed at the development of "teacher proof" materials. Most teachers embraced inquiry as an important rule of instruction (Goodlad, 1984 and Harms, 1977). However, classroom visits provided no evidence that it existed (Stake and Easley, 1978). The 1980s ushered in the inclusion of aspects of technology and environmental studies in school sciences to ensure that science was relevant to the individual in everyday living. The science, technology and society approach has arisen out of the need for relevance in education and in acknowledgement to a sense of social responsibility (Baez, 1986).

Meyer (1977) gave a useful analysis of the progressive change in goals. This analysis relates to secondary science education in Australian schools since the Second World War. According to him, in the decade following the Second World War, science courses were primarily concerned with the teaching of factual information, illustrating the technological applications of science principles. He named that era as the 'grab bag era' of science education. Around 1955 these loose collections of facts began to be replaced by 'big ideas' of science, breaking through the content barrier. He calls this era as 'big ideas era'. During this period there developed a curriculum reforms movement characterized by an emphasis on the nature and structure of science and by the use of teaching strategies that involve students actively in learning 'inquiry approaches' and 'discovery learning'. Contents continued to be selected for its capacity to contribute to 'big ideas'. But it was now organized and used in such a way that it illustrated the processes of science. Meyer calls this the breaking of the 'process barrier in the inquiry era'. He dates this at 1965. Meyer postulated the existence of a "value barrier" that began to be dented around 1975.

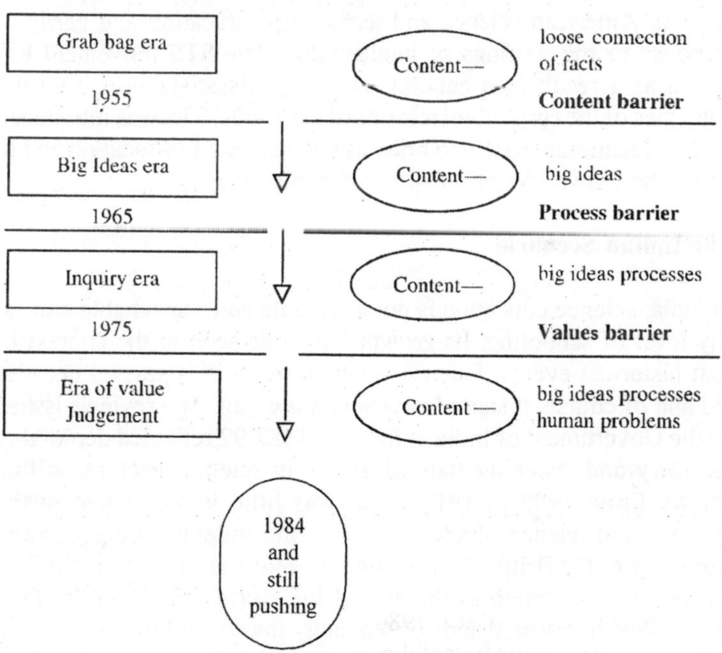

Fig. 1.2: Progressive Change in Science Goals

Source: Meyer, R. (1977). 'Science Curricula in Australia—A National View.' *Australian Science Teacher Journal*, 23(3), November, 90-97.

The 'era of the value judgment' that lay beyond this barrier was characterized as emphasizing the contribution of scientific knowledge and the processes of science to the amelioration of human and social problems. Meyer argued that the breakthrough of this barrier would effect a synthesis of content, process and values to bring about a 'confluence of understandings, attitudes, skills and insights to resolve conflicts between values'. Meyer's argument of the breakthrough of the 'value barrier' resulted in the inclusion of value aspects in the technological and environmental studies in school sciences in the 1980s.

It is in the concept of science, technology and society (STS) that there has been a discussion on the inculcation of processes and values to bring about a change in the understanding, attitudes, skills and insight of children. STS education is a term that has gained recent currency in the educational circles in America. It has arisen as a movement within American education partly in response to a personal crisis in American science and technology education and partly in response to the writings of intellectuals. The STS movement has grown as a result of a general feeling of dissatisfaction from the outcomes of the curriculum reforms of the 1960s. STS was introduced in the educational system to bring about meaning to education and to check the drawbacks of previous curricula reforms.

The Indian Scenario

In India, science education is not a separate and 'detachable unit' at any level of schooling. Its growth has to be seen in the context of past historical events. Science teaching in India is several decades old and of course, it started on a very slow foot. The reviews issued by the Government of India in the year 1887-92 reflected deplorably the sorry and shocking state of affairs in science teaching of that period. From 1905 to 1917, there was little improvement in the expansion of science education. India being under the protection and patronage of the British Empire did not enjoy the changes in the field of education as much as the rest of the world did. Then the First World War intervened and consequently interrupted this expansion. Science in India could not go through the processes of openly challenging beliefs and values. It developed as one kind of knowledge and a mere academic discipline. As a result of British dependency it had to miss, for historical reasons, the first Industrial Revolution.

Today, when the second industrial revolution has reached its peak in the West, India, like an adolescent in fever, is trying hard to achieve its past glory by recapturing its traditions of scholarship, original thinking and its great cultural heritage to participate fully in the march of science. Accordingly science becomes a priority area in education, both at the compulsory education level as well as at the level of specialization; however, considerable expansion in the secondary education occurred both in the rural and urban areas after the recommendations of the Calcutta University Commission.

Policy Interventions

Indian Parliamentary and Scientific Committee set up in August 1961 talked for the first time about the accelerated achievement in science, affairs of mankind and demand for increase in technically trained manpower and took up the study of science education in schools in 1962. Soon after this, in 1966, under the Chairmanship of Dr. Kothari a conference of science education was convened. This was held to plan an effective programme for the development of a total curriculum of science education at different stages. The Kothari Commission has pointed out that our science education is in bad shape and it will become worse if we fail to reckon with the explosion of knowledge.

As a follow up of the Kothari Commission Reports, the Ministry of Education and Social Welfare appointed an expert group in 1973 to develop curriculum for 10+2 pattern. New syllabi, text books and other materials were prepared but the text books evoked lot of criticism. As a result, in 1975, the National Council of Educational Research and Training (NCERT) attempted to give a concrete shape to the recommendations of the Education Commission (1964-66) and to the resolution of National Policy on Education (1986) by developing "The curriculum for Ten-Year School: A Framework". Due to various disparities this curriculum framework could not enhance science education in India.

Then came a period between 1984 and 1986 which is long remembered in the history of Indian education since innumerable activities in the form of conferences, seminars and group meetings were held to review the existing state of the art in education which led to the formulation of National Policy of Education (NPE, 1986).

One very important document "The National Curriculum for Primary and Secondary Education—A Framework" (1985) was brought out. Specific recommendations on science as a school subject in National Policy of Education (1986) emphasized on strengthening of science education so as to develop in the child well defined abilities and values such as the spirit of enquiry, creativity, objectivity, the courage to question and an aesthetic sensibility. Science education programmes were recommended to be designed to enable the learner to acquire problem solving and decision-making skills and to discover the relationship of science with health, agriculture, industry and other aspects of daily life.

The Programme of Action of NPE (1986) was formulated in 1992, which also emphasized on making content of science more organized to ensure that the approaches followed are in consonance with the objectives of science education as laid down in the policy. The NCERT revised the curriculum and syllabi in the light of the guidelines mentioned in the NPE (1986) and published a 'National Curriculum for Elementary and Secondary Education—A Framework, 1988'. In this document, the purpose of upper primary education (middle school education) is mentioned as 'at the upper primary stage (classVI-VIII) the learner is expected to consolidate and strengthen the abilities acquired at the primary stage'. The policy of integrating science at middle school level was clearly expressed in the document.

Technology Policy Statement (1983) also emphasized that use and development of science and technology must suit local needs and make an impact on ordinary citizens. In fact, in the recent years Yash Pal Committee Report (1993) also points out that what is required in our education is curriculum decentralization, localization and variety in curriculum content along with its relevance.

All this shows that India also has been making efforts to improve science education from time to time but they have failed to produce much impact. This could have been due to three major contextual constraints on school as given in the sixth publication of UNESCO, *New Trends in Integrated Science Teaching* (1990). The *first* is that most developing countries regardless of size and diversity have one science curriculum common to all schools and all children. *Secondly*, science education in most developing countries is doubly authoritarian both in its tradition and its pedagogy. *Thirdly*, the parents, children and teachers are very reluctant to accept curriculum change that appears to depart from 'elite' science which is traditional and academic science.

In India many changes have been brought in content, as well as methodology but still the stress on coverage of content, verification type laboratories, boring classes and rote learning continue to flourish. The objectives like creation of interest and curiosity, inculcation of an enquiring mind, promotion of problem solving attitudes and decision making powers, development of courage to question and growth of concern for other people and consequences have been talked about but never implemented. The entire planning is urban oriented and directed at the traditional learners who constitute less than three per cent of the total population.

From the above discussions, essential elements can be squarely framed for reformation of science education, i.e.

- To extend the curriculum to accommodate the changed knowledge brought about by technical and scientific advances.
- To make the curriculum available to wider sections of the population and see how to reduce it to the curriculum of elementary school.
- To enable the curriculum to serve the dual purpose of providing for a complete education for those not proceeding to higher education and a preparatory course for those who will be proceeding to higher education.
- To organize the courses in a social context rather than special disciplines.
- To have a balance of science and technology with an emphasis on inter-relationships between and among them, values and society at large.
- Development of courses around persistent social problems underlying health, environmental and technological aspects.

Why Science–Technology–Society (STS)

We live in an age where science and technology are not only integral parts of our contemporary culture but are also driving forces of our socio-economic development. As such, science and technology have a decisive role in everyday human endeavors in urban and rural areas alike. At the turn of the century our perception of the benefits of science and technology has changed considerably. When new vistas of progress through invention seemed to be opening up with the automobile, electric light, motion pictures, fast rail, road travel and later radio and the aeroplane, the benefits of science and technology seemed to be their over-riding characteristic. But now-a-days the number of negative examples of how society is linked with science and technology grows daily. Gas explosion that took lives of 500 slum dwellers in Mexico City and the leak of poison gas from the Union Carbide plant in Bhopal, India, which claimed the lives of over 1,000 men, women and children who died in agony in a situation in which medical assistance was lacking in both quantity and quality are two devastating negative examples. Both were disasters of a large order involving the three Ps: population, pollution and poverty. The fourth P — proliferation of nuclear weapons—could lead to a nuclear confrontation that would wipe out a large portion of the entire plant and animal population, including people. Science, technology and society approaches have arisen out of the need for relevance in education and in response to a sense of social responsibility.

Thus the rational use and application of science and technology can improve living standards and conditions. For this it is essential to have a population with a basic knowledge and an understanding of science and technology. Each and every individual should be able to recognize and evaluate the contribution science and technology are able to make to safeguard the existence of current and future generations. An important educational goal is not only the specialized knowledge of a few, but also the general knowledge required for everyone's every-day life.

Technology education which is incorporated in science education should be appropriate to the local requirements and technological achievements should be brought into focus in any implemented course or integrated curriculum approach. Teaching about the impact of technology on everyday life is essential. Pupil should be assisted in understanding the consequences of technology in society.

During the past few decades considerable advances in science and technology have resulted in important contributions to the quality of life in developed countries but basic problems associated with human sufferings of the majority of persons in the third world still remain. As a developing country our growth and labour situation, our standard of living and social security depend to a large extent on the quality of science and technology education for wide sectors of population.

It is, therefore, important to determine how to utilize recent developments in science and technology to ensure that meaningful benefits evolve in developing countries. It is viewed as a productive investment and an essential factor in the economic, social and technological development of the developing countries.

Our country confronts crisis every now and then, which has scientific overtones and implications that touch our national life deeply. The study of this crisis should be brought into the notice of our school children rationally and scientifically. The problem of the pollution of a river or of a certain region is intimately related to production problem in the region under consideration, and therefore to economic problems. Issues such as population growth, the use of urban and rural space resources, the use of medicine, sanitation, agriculture, all involve decisions that depend on a sound basis of information and knowledge. With the new threats of nuclear energy and diseases like AIDS, the ability of citizens in a democracy to engage responsibly in personal and public decision-making on issues which require some scientific understanding has become a matter of concern. Recent energy crisis, environmental pollution, ecology, population explosion and current shortage of food in the country created by drought and failure of rains in some parts and floods in other parts are examples. Such problems should be brought to the notice of school going children at an early stage, together with steps to combat them effectively. Importance of gobar gas, use of nuclear energy for development in the country, use of sunlight, wind power, geo-magnetic imbalances etc. should be taught at school level. Unfortunately, new crises and new hopes, which affect the life of both rural and urban population in this country intimately, do not find a place in the school curriculum.

There have been attempts to rejuvenate the science curriculum in India like the Kerala Shastra Sahitya Parishad (KSSP) which was started in 1964 (Zachariah and Sooryamoorthy, 1994) with the objectives to cultivate interest and awareness of modern science among the masses, to publish science books and periodicals in Malayalam, to organize meetings, discussions, science film shows and to assist other organizations working towards similar goals. Hoshangabad Science teaching programme in Madhya Pradesh, which was started in 1972 (Hoshangabad Vigyan, 1977) made another attempt. It is quite a novel programme in the history of Indian education of teaching science at middle school level based completely on 'inquiry method' and utilizing rural resources as instructional material. Though these programmes focus on local instructional material, technological and societal aspects were not emphasized upon. Moreover, the impact of these programmes was limited to local areas or to some extent to the state where they were being run. They did not have a national impact to be incorporated into the national curriculum.

The existing science material in particular needs careful examination. In the area of the curriculum where newspapers, books and television are a continual source of ideas, and text books are often out of date, a regular review of the relevance of existing materials needs to take place. Materials should be developed to increase instruction in the process skills of science and to make conscious efforts to improve the thinking skills of students. Teachers of conventional courses frequently "attempt" to cover the textbook and have little time for teaching the science process and thinking skills. There is need to stress the scientific method of acquiring knowledge rather than knowledge itself. If the student were equipped with the tool of the method of acquiring knowledge, the student would be in a position to learn throughout life according to his or her needs.

General science core curriculum is to be modified so that school students become aware of how the science they learn is related to the environment in which they live, it should be meant for everybody and not one that would produce only future scientists. In this context it is pertinent to quote Harms (1981) 'Can we shift our goals, programmes and practices from the current overwhelming emphasis on preparing all students to grapple successfully with science and technology in their everyday lives as well as to participate knowledgeably in the important science-related decisions our country will have to make in the future?' Hence the need to incorporate both science and technology in the teaching learning programme of all people from as early an age as possible and which is continued as an indispensable element of life-long education.

Yager (1990) gave the differences between the traditional science and STS based integrated science. The difference is summarized in Table 1.1 given below:

Table 1.1: Differences Between Traditional Science and STS Integrated Science

Traditional (Non-STS)	*STS (Integrated Trend)*
• Survey of major concepts found in standard textbooks	• Identification of problems with local interest/impact
• Use of laboratories and activities suggested in textbooks and lab manual	• Use of local resources to locate information that can be used in problem solving
• Passive involvement of students assimilating information provided by teacher and textbook	• Active involvement of students in seeking information for use
• Learning being contained in the classroom for a series of periods over the school year	• Teaching going beyond a given series of class sessions, a given meeting room
• A focus on information is proclaimed important for students to master	• A focus upon personal impact–often starting with student curiosity
• A view that content is the information included and explained in textbooks and lecture	• A view that content is not something that merely exists for student mastery simply because it is in print
• Learning focusing on current explanations	• Learning with a focus on the future
• Learning is principally for testing	• Learning occurs because of activity
• Retention is very short-lived	• Students who learn by experience retain it
• Students see science processes as skills scientists possess	• Students see science processes as skills they need and can use
• Students see processes as something to practise as a course requirement	• Students see processes as skills they need to refine and develop
• Students do not understand teacher concerns for process, especially as they rarely affect course grades	• Students readily see the relationship of science processes to their own actions
• Students' interest decline at a particular grade level and across grade levels	• Student interest increases in specific courses and from grade to grade
• Coursework seems to decrease curiosity	• Students become more curious as they learn and do more
• Students see teacher as a purveyor of information	• Students see teacher as a resource and guide
• Students decline in their ability to question: the question they ask is often ignored because they do not fit into the course outline	• Students ask more questions, such questions are asked to develop STS activities and materials

(*Contd.*)

Table 1.1 (*Contd.*)

Traditional (Non-STS)	*STS (Integrated Trend)*
• Student rarely asks unique questions	• Students frequently ask unique questions that excite their own interest
• Students are ineffective in identifying possible causes and effects in specific situations	• Students are skilful in suggesting possible causes and effects of certain observations and actions
• Students have few original ideas	• Students seem to effervesce with ideas
• Students see no value or use of their study to life	• Students can relate their study to their daily lives
• Students see no value in their study for resolving current societal problems	• Students become involved in resolving societal issues
• Practice of basic process skills—but little attention to them in terms of evaluation	• A de-emphasis on process skills—just because they represent glamorized skills of practising scientists
• No attention to career awareness other than an occasional reference to a scientist and his/her discoveries	• A focus on career awareness, especially careers that relate to science and technology and not merely to those related to scientific research and engineering
• Students concentrate on problems provided by teacher and textbook	• Students perform in citizenship roles as they attempt to resolve issues they have identified.
• Learning occurring only in the confines of the class room	• Study being visible throughout an institution and in a specific community
• Science being a study of information where teachers determine the amount that students acquire	• Science being an experience students are encouraged to have

Source: Yager, R.E. (1990). *A general view of Trends in Integrated Science Teaching*. UNESCO Publication.

This contrast shows major differences between the STS and non-STS science. The differences emphasize the advantages and the power of STS. These include students' growth in the processes of science, activities on the interest of the student and relevant to their lives. It builds on local resources, focuses on career awareness and helps students in performing citizenship roles. Learning in STS occurs with the help of activities and with a focus on the future; students show interest in the course and are curious to learn more. Teacher acts as a resource person or guide; students ask unique questions and suggest possible causes and effects of certain observations and actions. Students relate their study to their daily life and in resolving societal issues. The traditional science, that is non-STS science, on the other hand covers major concepts found in standard textbooks. The activities are as suggested in textbooks and laboratory manuals. Students are passively involved in learning; there is no attention to career awareness, students concentrate on problems provided by the teacher. Learning is focused on current explanations and retention is short-lived. Students interest declines and teacher is seen as a purveyor of information. The ability of the students to ask unique questions declines, very few of them have original ideas and they see no use or value of the knowledge in their lives.

Hall *et al.* (1983) in *Teaching Science, Technology and Society in the Junior High School* cite eight main reasons for teaching STS. These are:

- A recognition that science is a part of our culture and as such the part it plays should be understood;
- A belief in the importance of an informed citizenry, particularly in a technological age;
- A concern for the *quality of life* particularly as scientific and technological developments impact on this quality;
- A previous neglect of the values that underpin developments in science and technology;
- The rise of a concern among scientists for the social responsibility of their work;
- A recognition of the social context in which science is practised and of their mutual influences;
- Developments in core curricula that point to STS as a central part of such a core;
- A desire for relevance in secondary science curricula.

The Focus of STS

The new curriculum based on science technology and society (STS) should incorporate and emphasize the inter-relationships of science, technology and society to produce educated laymen regardless of the diverse nature of their future careers. STS stands for the ability of the students to integrate their personal understanding of the natural world with both the man-made world and social world of day-to-day experience. It will enable young students to develop skills and acquire knowledge which will produce a better way of life for them in rural life, which will make it possible to lead the students to the discovery of knowledge. It would aim at raising the level of scientific knowledge, skills and attitude of the students to allow them to be more productive in their home environment.

STS is designed to make science curriculum relevant to the students so that science is seen as interacting with society. It is not just a one-way street, it is not just that science influences society but there is feedback from the society also. Students not only understand science but also become critical thinkers. STS is how to bridge the gap from what the kids learn in the classroom as science and where it applies in their everyday lives.

STS is concerned with the application and issues of science in society. Such issues and application should be integrated in science courses effectively. This is in contrast to a conventional science course, which might start with a particular scientific fact or theory such as the properties of acids or the laws of motion. With the integration of the STS themes in the science courses the starting point might be an issue, such as 'Do we need food additives?' 'How to solve the problem of traffic safety, alcohol abuse and heart diseases?' 'Can we build a dam without harming the environment?' 'How do we solve the problem of pollution?' 'What can be the alternative source of energy?'etc. From these starting points, relevant scientific facts and theories can be developed as and when they are needed for an understanding of the issues concerned. Such integrated science courses aim to teach pupils science that will be useful in their future, whether or not they choose to study science further. Teaching of such courses requires a variety of teaching and learning activities which reflect a wide range of skills and processes involved in learning about science in the real world.

Science-technology-society instruction refers, therefore, to teaching about natural phenomena in a manner that connects science with the technological and social worlds of the student.

Throughout the 1980s, STS programmes were developed in many countries. Technology here was not necessarily subservient to science, although in fact few of the earlier courses acknowledged technology as more than 'applied science', inseparable from and dependent upon it. The focus was on societal issues and topics with a strong science and technology dimension and the main contribution to the aims of an autonomous technology education was through 'technological awareness', i.e. awareness of the personal, social, moral, economic and environmental implication of technological development. Later attempts to provide a conceptual framework for STS courses have, however, given greater prominence to understanding the nature of technology and its interactions with science and society (Aikenhead, 1987; Bybee, 1987 and Layton 1988).

STS studies at secondary levels have been carried out in Great Britain for over thirty years (Solomon 1980, 1988). The Northern Examining Association (1988) prepared its syllabus for STS at secondary level. Science through Science, Technology and Society Project (1985) also developed a set of guidelines for instructional materials' evaluation in STS.

A review of the literature reveals the importance of an STS curriculum to meet the needs of a great number of students in industrialized and developing countries (Bybee *et al*, 1980; Eikelhof *et al.*, 1982; Hall *et al.*, 1983; Hurd, 1975; Lewis, 1978; Morland *et al.*, 1981; McConnell, 1982).

Research evidence is beginning to build up which suggests that compared to students in more traditional classes, US students who have STS framework learn just as many science concepts, in some instances more, and retain these concepts for a longer period of time (Blunck and Yager, 1990; McComas, 1989; Myers, 1988; Yager and Penick, 1988). They demonstrate more of an ability to apply knowledge than traditionally taught students, as measured by: the use of information in new settings; the understanding of phenomena experienced in new settings; the identification of appropriate questions to ask in problem solving; the choice of information to solve problems; and the choice of appropriate action based on new information. There are also some evidencer of enhanced development of process skills with the STS approach (Yager, 1990).

Infusion of STS in Textbooks

Most teachers, students and parents accept textbooks as central and the only source of nearly all information and as the framework from which all science is to be experienced. Most teachers use textbooks extensively following them page by page. Thus, in-depth information concerning STS needs to be added to the content. Science and technology related issues or problems need to be identified and addressed. Inclusion of such themes improves student growth in the affective domain (McComas, 1989; Yager, 1990) and hopefully with such a focus in science text books, students will sustain their interests in current events, discern inter-relationship on their own, care enough to participate in making changes in their own lives and become actively involved in efforts to improve society.

At the upper primary level, the curriculum needs to develop a more utilitarian focus. It is probably sensible to begin by gradually adding relevant societal content to existing courses, since this strategy can be implemented relatively quickly, and with less expense than a major requisite of the syllabus. Even for teachers it is less threatening. It should not seem to the students as a separate subject, divorced from anything that they are learning. Thus STS themes should be introduced into the already existing science courses.

. The issue on infusion of STS into the already existing courses was also discussed in *International Organization for Science and Technology Education* (IOSTE) symposium on World Trends in Science and Technology Education held at Kiel, Germany in 1987. The working group discussed separate STS courses *versus* infusion (integrated or "piggy-backed") of STS into existing courses: "While separate STS courses offer systematic STS instruction balancing basic science content with its technological and social milieu, in most school systems the number of academically oriented students enrolling in STS courses is limited, and thus non-academic students are often designated for such courses. On the other hand, integration of STS into existing courses will reach a wider group of students and will not explicitly compete with the *status quo*, however, implicitly, the STS content will compete with other day-to-day demands and routines of teaching, and consequently the STS aspects will tend to be shelved easily, due to their infusion status. The working group suggested that "the infusion or integration of STS into existing course material would likely be the preferred way to proceed. Research on both approaches is needed to illuminate the strengths and weaknesses of each type of material".

The approach of infusing STS issues into an existing science curriculum was specifically chosen because it appeared to be the most viable one under the prevailing socio-economic conditions in a developing country like India.

Why at Upper Primary Stage

Keeping in view the recommendations given in NPE (1986) it is our national commitment to make the curriculum of elementary stage more need–based and relevant to life, and teaching/learning situations to be selected from both rural and urban areas. It is especially essential to link science of upper primary classes with the constitutional commitments, different priority areas of national development (agriculture, communication, transport, textile, chemicals etc.) and certain thrust areas (energy, environment, social forestry, wild life management, industry). At the upper primary stage only these thrust areas, or in turn we can say STS objectives, can reach the masses, as education is compulsory and universal upto the age of fourteen. After this stage the education is neither compulsory nor universal.

It has generally been seen that when students come to the upper primary stage, i.e. from Class VI to VIII, they have a negative attitude towards science and mathematics. A positive attitude can be developed more if the concepts taught are all the more useful to them in their daily lives and are related to society. At this stage the abilities acquired at the primary stage are all consolidated and strengthened besides helping the learner in understanding certain simple chemical, physical and biological principles. The child is also helped to develop capabilities of using processes of science in daily life. The learner is also helped in acquiring skills of manipulation, designing of simple experiments, appreciating the inter-relationship of science and technology and their relationships with society and in developing proper social and moral values.

As an element in the upper primary stage curriculum science, technology and society provides a method for fulfilment of objectives, solution of problems, enrichment of individual subjects and a link relating them to life and to each other.

The science content of upper primary school curriculum is more closely specified and organized than that of the primary level and so organized as to relate to major science concepts. Moreover this stage is the terminal stage of education for many students. A large number of learners enters public life after completing upper primary school education and so they must be taught some science which is required to understand the fast changing scientific world and to appreciate the achievements of science, which will help them to develop such attitudes and values as will make them capable of making wise decisions in future life as well as develop abilities to participate actively in national building.

The conceptual understanding of content area at upper primary stage has some additional significance because it is a foundation stage for science education and for subsequent build up it is necessary to have a stronger and sounder foundation. It becomes very difficult and sometimes even impossible to overcome the wrong concepts formed at the foundation stage. The upper primary stage is the stage where science education starts in a formal way, before which they study environmental studies. As such, students at this stage should be armed with the dangers and follies of our modern age. If STS were introduced at a stage later than upper primary level, fifty per cent of the students would be unaware of the real applications of science and technology because they would not have exposure to STS education. Thus it was thought right to conduct the research at the upper primary stage.

Importance of STS for Rural Areas

According to the Sixth All India Education Survey (NCERT, 1998), for 5,87,247 inhabited villages there are a total of 5,11,849 primary schools, 127,863 middle schools, 48,262 secondary schools and 11,642 higher secondary schools. Going by pure averages, this leaves 13 per cent villages unserved by a primary school, 78 per cent without a middle school, 92 per cent without a high school and 98 per cent without a higher secondary school. At the present juncture less than 25 per cent villages have a middle school and bulk of our children in India go only as far as primary level and at best up to class VIII. In the foreseeable future, elementary education is the maximum to which most of the population, especially rural children would have access. This implies that for majority of our children and adolescents, primary/ upper primary education will be the terminal stage. All of these young persons would enter life, marry, have children, work for their living, set up small business or enter wage work. They would be the future aspirants and members of the Panchayats and local bodies and as alert citizens and human beings, fight social evils and man-made/ natural disasters, guarding their fundamental rights and fulfilling their duties. In a world fraught with deceit, drugs, violence and diseases, these young ones have not only to learn to fight castesism, consumerism, partisan values etc. but also must develop the capacity to harness science and technology for human development, balancing the twin principles of growth and justice.

We see that most of the rural children get eliminated from the qualifying rounds. They have very slender chances of completing ten to twelve years of schooling, which would entitle them to higher general and technical education. We have an excellent national curriculum framework, which gives broad guidelines with enough room for local adaptation. However, curriculum transaction is incomplete, fragmented and truncated. There is a need for curriculum improvement, especially in science for the rural population of students studying in upper primary schools in India. There is need to lay emphasis on the STS oriented science curriculum pertaining to the needs of the whole state which would help the students to have insight into the developmental changes occurring in the city as well as making the best use of the knowledge gained in their immediate environment.

Most of the changes and educational movements are restricted to urban areas and less importance is given to changes in the educational system of rural schools, but India lives in its villages. It is there that our producers live, voters live, and the poor and the illiterate live. It is the villages that hold the country's problems. There is need to rebuild the half-a-million villages and teeming millions inhabiting these villages of ending the sufferings.

An understanding of the utility of science is especially important in the rural areas. For example, the student who becomes a farmer, or who works in an agricultural community, needs to be able to apply both traditional and modern ways of enhancing productivity. The natural cycles need to be understood, as well as ways to enhance crop yields safely and cost effectively using synthetic fertilizers and pesticides appropriately. An ability to maintain and repair equipment is also useful. The health worker in the village needs to understand basic biology including the life cycle of various parasites and other vectors of human disease. Village woman who understands the working of her reproductive system in biological terms may use birth control devices more successfully.

Appropriate technology can have beneficial effects on relieving poverty in rural Indian environment. According to the planners the major objectives of rural development are removal of poverty and unemployment. These objectives can only be achieved if the rural population has a positive acceptance and attitude towards the appropriate application of technological solutions.

In the light of the above discussions the present study was conducted keeping all these aspects in mind. Though many researchers have been conducting on science education in India, the STS aspect of science education has remained an area neglected by Indian researchers. The researcher in the present study has tried to evolve

STS themes pertinent in the Indian context and has worked on the aspect of infusion of these themes in the present science textbooks of the Haryana Education Board. Haryana was particularly chosen for the study as the state is representative of rural backward villages where scientific literate people are needed to overcome the age-old superstitions and practices. The state has made rapid development in infrastructure, high agricultural growth and the white revolution, which made the state prosperous, but despite the material progress, the state continues to register poor performance on all well-established social indicators. Infusion of the STS theme in the already existing curriculum of upper primary level may be the most realistic approach for curriculum framers to keep the technological and social aspects in mind while framing the upper primary science curriculum in future.

2

Theoretical Framework

Introduction

Technology applied to science has been taught for as long as there has been science education but STS presents the structure of science and technology in relation to society. This new approach to science has arisen because of the capacity of science to provide wealth and health for society, the will to bring craftsmen and the general public towards an understanding of science, to overcome the shock of negative effects of science in warfare and in the environment, and for the need to assign values and responsibility with science.

This chapter covers the following aspects (i) definitions of science, technology and society, (ii) multiple connections within science, technology and society, (iii) inception of STS, (iv) models on conceptual frameworks, (v) models on curricular frameworks and (vi) specific projects dealing with STS. The literature under this chapter is limited to their availability during the review of related literature. Some of them are described briefly, as a detailed account of them could not be procured.

The definition of science, technology and society independently and with interactions of each other helps to understand the meaning of STS. The inception of STS discussed in this chapter deals with the start of STS education in the world. It helps us to understand the need felt for a change in shift of curriculum focus.

The models of conceptual framework on STS give an insight into the theoretical formulation of the STS curriculum. The conceptual framework clarifies the understanding of how to go about the STS issues. They fulfil several important developmental and pedagogical criteria for the translation of the concepts, processes and an understanding of the curriculum. It helps in conceptualizing the various components of STS. The conceptual framework includes the primary area of focus, various themes, issues and content area of STS. Three models of conceptual framework given by Mac-Cleave Frazier and Murray (1984), Bybee (1987) and Hickman, Patrick and Bybee (1987) have been discussed in this chapter.

The curriculum framework includes the various aspects involved in formulation of STS curriculum. The models of curriculum framework provide specific guidelines, strategies and evaluation techniques required for better implementation of STS. The models discussed in this chapter give an insight into 'doing' of STS, based on the conceptual framework models. A sound curriculum framework is based on a sound conceptual framework. In response to the pressing needs of modern society the curriculum framework offers a new educational orientation, which would compensate and complement the formal traditional science education. Framework involves the major phases of activity involved in the planning of the curriculum. Three models of curricular framework given by Driver and Oldham (1986), Waks (1992) and Cheek (1992) have been discussed in this chapter.

The specific curricular projects undertaken by various researchers deal with the implementation of the curricular packages. Evaluation of the effectiveness of the curricular package helps in promoting students' learning and portrays the success with the material produced. Projects are the efforts put in by various researchers and educational groups to judge the practical feasibility of the curriculum. Projects discussed under this chapter were taken up to organize STS instruction, STS topics to be included and development of the STS modules. Nine projects have been discussed under this heading. Lastly a comparison of guidelines and curricular framework of STS education given by Cheek (1992) have been discussed.

Defining STS

The definition of the term STS has to be completely understood to have a clear consensus about its various aspects. The persons active in this field have provided some working definitions, which are as follows.

Aikenhead (1986) suggested the following working definitions of science and technology:

Science: Is a social institution possessing disciplined processes, which utilize knowledge and technique to achieve conceptual, material and social goals.

Technology: Applies scientific and other domains of knowledge/ resources to practical problems that arise from human purposes, guided and constrained by organizational systems, people, hardware and nature.

Bybee (1986) gives a different set of definitions for "Science", "Technology" and "Society".

Science: A systematic, objective search for understandings of the natural and human world. Science is characterized by use of an empirical approach, statements of generality (laws, principles and theories) and testing to confirm, refute or modify knowledge about natural phenomena.

Technology: The applications of scientific knowledge to solve practical problems to achieve goals. A body of knowledge developed by a culture that provides methods or means to control the environment, extract resources, produce goods and services and improve the quality of life.

Society: The collective interactions of human beings at local, regional, national and global levels. Human groups whose members are united by mutual interests, distinctive relationships, shared institutions and common culture. The human sets in which scientific and technological enterprise operates.

It is clear that multiple connections of these terms are possible and even useful. The above definitions adequately convey the fundamental aspects of science, technology and society as separate constructs. Science, technology and society (STS) education, then, is concerned with the manner in which these constructs, as represented by institutions, ideas, and people interact with one another in a dynamic manner. There are six kinds of interactions among them, as given below by McKenzie (1983).

Science on Technology: Some new technologies arise directly from advances in scientific knowledge and some from modifications to older technologies arising out of the application of science.

Technology on Science: Development in technology can lead to new or improved instrumentation and techniques that lead to major advances in science.

Science on Society: Science has had an influence not only at a practical level on the quality of our lives, but also on our world-views and on our ideas about knowledge.

Society on Science: Scientific knowledge is generated in a social context which influences the research undertaken, the reception of the research findings, as well as the content and structure of the theories themselves.

Technology on Society: Modern technology has contributed in dramatic ways to the quality of our lives with consequences both beneficial and detrimental to mankind.

Society on Technology: Social norms, values and political processes influence the direction of technological change (Lowe 1982).

These interactions can be further depicted by the Fig. 2.1 given by Glen Aikenhead in one of the group reports on Science and Technology Education and the Quality of Life presented at Kiel, Germany, 1987.

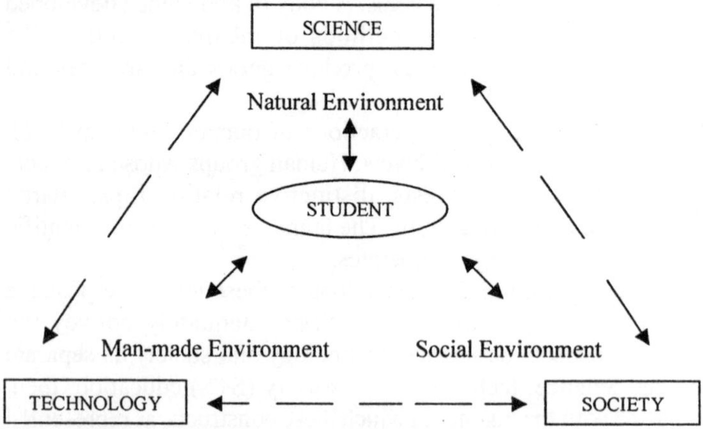

Fig. 2.1: Interactions Between Science-Technology-Society

Source: Aikenhead, G.S. (1987). Group Reports. In Riquarts, K.,(ed.)(1988) *Science and Technology Education and the Quality of Life*, Institute for Science Education (IPN), Germany.

As indicated in Fig. 2.1 (solid arrow), students integrate their personal understanding of the natural world with both the man-made world and social world of day-to-day experience. The study of the natural world is science. The study of the man-made world is technology. And social world is the society. Science-technology-society science inclination refers, therefore, to teaching about natural phenomena in a manner that connects science with the technological and social worlds of the student (the broken arrows in Fig. 2.1).

In 1988 the National Science Teacher Association (USA) decided to define STS broadly as a means of defining current reform efforts. The Special Task Force established for this purpose defines STS in the following manner; STS is a term applied to the latest efforts to provide a real world context for the study of science and for the pursuit of science itself. It is a term that elevates science education rhetoric to a position beyond curriculum and the ensuing debate about the scope and sequence of basic concepts and process skills. STS includes the whole spectrum of critical incidents in the education process, including goals, curriculum, instructional strategies, evaluation and teacher preparation/performance. STS means focusing upon current issues and attempts at their resolution as the best way of preparing people for current and future citizenship roles. This means identifying local, regional, national, international problems with students, planning for individual and group activities which address them, and moving to actions designed to resolve the issues investigated. STS provides a setting and a reason for considering basic science and technology concepts and processes. It means determining ways that these basic ideas and skills can be seen as useful. STS means focusing on real world problems instead of starting with concepts and processes which teachers and curriculum developers argue, in terms of future usefulness to students.

The Science and Society Committee of the National Council for the Social Studies (USA, 1990) defines STS Education as the understanding of how science and technology shape and get shaped by society, the problems and opportunities they create, and how citizens can relate most effectively to them.

Inception of STS Education

First World War is referred to as 'Chemists War' and the Second World War as the 'Physicists War', but the repercussions on science of the Second World War differed in kind, as well as in discipline, from those of the first. The blame for gas warfare in the First World War was not put on the scientists but on the industry who produced it. After the Second World War, the situation was completely different as all the scientists from all over Europe flocked to Los Alamos to work, quite knowingly, on what they called "our bomb". The effects of actually dropping the atomic bomb on Japan shocked them. They realized the after effects of their scientific challenge as it got carried away by intellectual problem solving and ignored morality. The oil crisis of 1973 led to increased concern for the conservation of non-renewable resources. After the war there was a new organization of scientists who worked on the need for 'citizen science' in education, i.e. public awakening to the environmental effects of new technologies which first made the need felt strongly in the United States. In the 1950s and 1960s the extent of pollution had begun to shock the whole nation, the Great Lakes had been badly polluted by industry with the result that one had become completely lifeless. Environmental groups demanded to know more about the activities of science.

In Britain this was reflected in the UK's Freedom of Information Act in 1967. By the 1970s there was expression of a new reason and function for science education. Scientists were eager to solve problems of the developing world. They focused attention on the new technologies of alternative and nuclear power. Michael Young (1971) and later Malcolm Skilbeck (1984) examined the educational debate and purpose of education for social reconstruct and cultural transmission. Both these educational arguments about equality of opportunity within the curriculum and about citizen empowerment were tantamount to a new invitation. STS, which emphasises on social responsibility, was needed inside the curriculum to complement the more traditional approach to science education. The first mainstream school science course to take up the challenge to include a strong STS component was SCISP (School Council Integrated Science Project, 1973). It taught directly about economics and industry as well as other STS themes.

The SISCON (Science in a Social Context) Project had been started in Britain in 1970 to promote STS teaching, in universities and polytechnics. Being confined to degree courses, the intention of STS of bringing science to a wide audience so that the public could be empowered to take part in public and political issues could not be realized. For that it needed to go into mainstream schools. In 1978, SISCON-in-schools was started which focussed on how STS should show pupils about the fallible nature of science and how STS education includes discussion of democratic action.

Societies of different countries differ and so do their education, systems and the economic situation of the society concerned. In USA during the 1960s, there were several reports on the nation's science education. The National Science Foundation wanted to increase the quality of science education so that there would be more scientists.

The government wanted more trained scientists to do scientific jobs so that the nation could match the Russians in new nuclear arms (or later to match the Japanese in industrial technology). Twenty years later, reports highlighted general confusion about educational goals for science education and linked this with public apathy towards science (Yager, 1980). From this time, STS began to figure prominently in discussion and in curricula (McConnell, 1982). A number of events in the world probably triggered these movements. The disillusionment caused by the tragedy of the Vietnam War compelled rethinking of all national curricular developments, and focused on the negative aspect of science and technology (Yager, 1984). The 1980s ushered in the inclusion of aspects of technology and environmental studies in school science to ensure that science was relevant to the individual in everyday living.

In Canada STS course appeared as early as 1971. In *Science—A Way of Knowing,* Glen Aikenhead (1975) produced a series of classroom materials for teaching the philosophical and social aspects of science within high schools, which were in advance of those from any other country.

Holland had a special national need for STS in the 1970s which ensured that their new school courses had matching objectives. Holland went further by setting aside a period of eight years for public education and consultation about nuclear power. It was during this time and on the premise of public participation in decision-making, that the early Dutch STS courses began (Boeker, 1979).

Rustum Roy, Director of the Science/Technology/Society (STS) Programme at Penn State University has proclaimed that STS is a megatrend in science education today (Roy 1983, 1985a, 1985b). The National Science Teachers Association (NSTA, 1982) has proclaimed STS as the focus for school science. The NSTA has published two year books which consider STS parameters for exemplary science instruction; it has produced two monographs dealing with issues to prepare science teachers for STS instruction. STS is also central to reform efforts in social studies education. One of the major task forces of the National Council for the Social Studies is on STS. The National Social Science Association devoted its 1987 meeting to STS issues. The Social Science Consortium has produced a definitive rationale and framework for STS education. STS is becoming a central thing in the social studies, emphasizing the centrality of science and technology to current culture.

State-wise STS efforts got underway in Iowa in 1983 as part of the National Chautauqua. In-service Programme operated by NSTA during the 1983-86 interim. The Iowa programme was the only one that focused on STS materials and strategies. The Chautauqua Model was seen as the one that could affect teacher attitude and behaviour.

STS studies at secondary school levels have been available in Great Britain for over twenty-five years (Solomon, 1980b, 1988b). The Northern Examining Association (1988), in preparing its syllabus for STS, specified the following five aims for the course in terms of opportunities for students:

- Explore some instances in which sciences, technologies and societies interact, based on the concepts and themes of the common core of this syllabus.
- Develop an understanding of arguments representing widely diverse priorities and viewpoints including cultural, moral, religious and other values and formulate their own reasoned responses to such arguments.
- Make reasoned choices of any decision about appropriate technologies based on the recognition of the complex inter-disciplinary nature of the problems and of the diversity of viewpoints.
- Develop an interest in and concern for the global environment.
- Develop and maintain an interest in current topical issues in science, technology and society.

The syllabus then goes on to specify objectives which will be assessed in three areas: (1) knowledge and understanding, (2) comprehension and application, and (3) evaluation.

The call for STS began to make itself felt in the developing countries also. Initially there was attempt to follow the school science programmes in lines with Western courses. But borrowed educational themes and resources were often quite grossly at odds with local natural examples, local culture and local needs. No discussion of meaningful STS issues could flourish within such imported syllabuses.

Conceptual Framework

Conceptualizing the various interactive components necessary for STS education is difficult. STS issues are by their very nature multi-disciplinary, controversial and not susceptible to final solutions. A conceptual framework clarifies the understanding of how to go about STS issues. It fulfils several important developmental and pedagogical criterion for the translation of concepts, processes, and understanding into curriculum programmes and instructional practices.

Conceptual Framework for Teaching STS Studies (Mac-Cleave Frazier and Murray, 1984)

Mac-Cleave Frazier and Murray (1984) have developed a conceptual framework for teaching home economics. This framework has been substantially revised for STS education and rendered more interactive than what the original authors intended in their Model (Cheek, 1992). Fig. 2.2 shows two panels (Integrating Perspective and Systems of Action), that are processes that mediate between the other panels. They determine selection and sequencing of items from adjacent panels. A quality STS education approach may be expected to include an emphasis upon the items within these two panels. The dynamic aspects of the model are reinforced by the use of arrows proceeding in either direction. This means that in STS education we can start either at the far right (letting an STS issue be our starting point) or at far left panel (i.e. from a discipline) in conceptualizing our ideas. Panels A, C and E can be interchanged while panels B and D remain static. We can see how this conceptual model aids thinking about STS education by considering the following energy resources example. Let us begin with the far right panel.

Area of Focus

The practical problem that confronts us as human beings is the non-renewable nature of many of our current energy resources. Even so called renewable resources have finite limits in the amount of energy we can usefully and efficiently extract from them given spatial and other limitations.

Systems of Action

Scientific/Investigative Skills: Calculation of a unit of energy using experimental procedures, laboratory investigation of the limits within systems, investigation of energy sources utilized within the local area.

Special/Political Action Skills: Writing letters to the local newspapers concerning energy policy, contacting the congressional and state legislative representatives regarding current technologies under review at national or state levels and communicating student opinions on energy policy considerations to local community leaders and utility companies.

Policy Making Skills: Understanding and experience in the roles played by debate, communication, consensus, data, expert opinion, lay opinion etc.

Critical Evaluation Skills: Tradeoffs, reliability of the data in question regarding energy reserves and usage, implicit values regarding energy resources and usage, and implicit value and their roles in the discussion concerning energy and resources etc.

Knowledge in the field of STS

There are a number of different ways in which students could be led through an investigation of energy and resources.

Content Areas of STS: Policy-making and energy, history of energy usage etc.

Various Organizers: Limitations (extraction, utilization, thermodynamics etc.), waste (heat, materials, products, etc.) and a host of others.

Topics: Nuclear energy, solar energy, geothermal energy, coal etc.

Concepts: Energy production, energy consumption, energy exploration etc.

Generalizations: There is no such thing as a free lunch; no man is an island and neither is a society; what is here today is gone tomorrow; we are borrowing on our children's future etc.

Themes: Better utilization of resources, more equitable distribution of resources, future developments in energy resources, past lessons from energy utilization, etc.

Issues: Should nuclear power plants be closed down? Should we return to coal as a main source of energy? Should the federal government massively fund research into solar energy? and so on.

Integrating Perspectives from STS

Synthesis: What does data from the respective disciplines bear upon the use of energy resources? Library skills, interview skills, media utilization for information, computer network and on line services, information on energy resources etc.

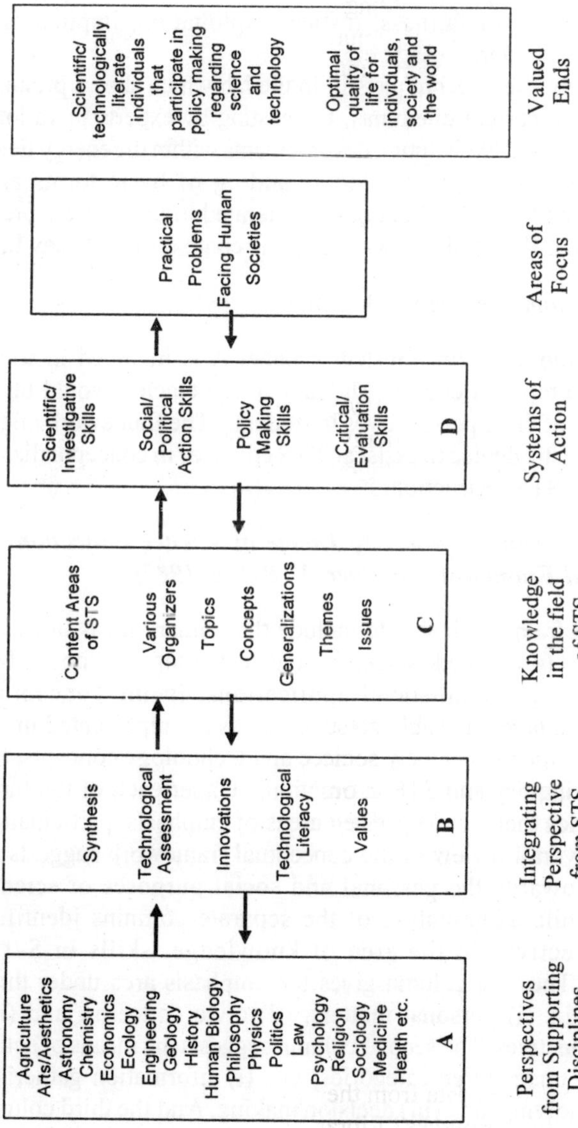

Fig. 2.2: A Conceptual Framework for Teaching STS Studies

Source: Cheek, D.W. (1992). *Constructive About Science, Technology and Society Education.* State University of New York Press.

Technological Assessment: Efficiency of current technologies, risks associated with current technologies, scope and economic linkages of the energy business etc.

Values: Issues of fairness, justice, exploitation, implicit and explicit values in energy usage etc.

Innovations: New technique still in the lab which shows promise in helping solve current dilemmas, forecasting by experts in various disciplines regarding likely future developments within the energy field.

Technological Literacy: Understanding of basic terms and concepts related to the ways energy is generated, transmitted, stored, used, measured, financed and wasted in the context of everyday life.

Perspectives from Supporting Disciplines

Everyone of the disciplines tested in panel A is involved in some ways with the topic of energy. The task of the teacher would be to develop some of these perspectives in students. The framework might be useful heuristic device to assist an STS educator in conceptualizing and acquiring STS instruction.

The Science-Technology-Society Theme in Science Education: A Conceptual Framework (Rodger W. Bybee, 1987)

A conceptual framework should include the primary areas in which objectives are formulated: scientific and technological knowledge, processes and personal/social applications. In the conceptual framework displayed in Table 2.1 those areas are represented in the three essential themes, namely science and technology concepts, the processes of inquiry and STS interactions. Under each of the three main categories there are suggested areas of emphasis, participation and study. Overall review of the conceptual framework suggests an orientation towards the personal and social purposes of science education while examination of the separate columns identifies specific objectives in the area of knowledge, skills in S-T-S interactions. The first column gives the emphasis area under three categories, viz. (i) personal matters, (ii) civic concerns, and (iii) cultural perspectives. The second column suggests the means of active participation under three categories, viz. (i) information gathering, (ii) problem solving, and (iii) decision making. And the third column gives what should be studied under three categories, viz. (i) local issues, (ii) public policies, and (iii) global problems.

Table 2.1: Conceptual Framework for Knowledge, Skills, and Values of Scientific and Technological Literacy

Acquisition of Knowledge	*Utilization of* Learning skills	*Development of* Values and ideas
Related to Science and Technology	*Raised on* Scientific and Technological Inquiry	*About* Science & Technology in Society
In three areas of emphasis Personal Matters Civic Concern Cultural	*By means of active participation in* Information Gathering Problem Solving Decision Making	*Through study of* Local Issues Public Policies Global Problems

The acquisition of knowledge related to science and technology is fundamental to science education. Concepts that might be included in the first column are those that have been described as fundamental. Some examples of concepts to be included in STS programme are listed in Table 2.2. The areas of emphasis listed provide the contextual orientation for the topics and activities. Table 2.3 displays inquiry skills for STS programmes. These skills are more specific examples of the general areas of participation listed in Table 2.1. Finally, Table 2.4 lists the fundamental science technology society interactions. Tables 2.2, 2.3 and 2.4 are given here to provide accurate interpretation of the conceptual framework. Progression down the columns leads from the personal to cultural, information gathering to decision-making, and local to global perspective. Generally the framework goes from simple to complex, concrete to abstract and immediate to past and future perspectives.

Table 2.2: Unifying Concepts for the Science-Technology-Society (S-T-S) Theme

System and Sub-systems
Organization and Identity
Hierarchy and Diversity
Interaction and Change
Growth and Cycles
Patterns and Processes
Probability and Prediction
Conservatrion and Degradation
Adaptation and Limitation
Equilibrium and Sustainability

Table 2.3: Inquiry Skills Based on Science and Technology

Questioning and Searching
Observing and Organizing
Measuring and Classifying
Comparing and Conserving
Analyzing and Synthesizing
Identifying and Describing
Hypothesizing and Predicting
Separating and Controlling
Exploring and Evaluating
Deciding and Acting

Table 2.4: Themes of Science-Technology-Society Interaction

Science
Technology
Society
Science and Technology
Science and Society
Technology and Science
Technology and Society
Society and Science
Society and Technology
Science and Technology in Society

Source: Bybee, R.W. (1987), Science Education and the Science-Technology-Society (STS) Theme, *Science Education*, 71(5), pp. 667-683.

Five Questions as a Guide to Selecting Appropriate Content for STS Course (Hickman, Patrick and Bybee, 1987)

Selection of the content that should appear in an STS course or unit poses many of the same problems as selecting content for any other course or unit one wishes to teach. Hickman, Patrick and Bybee (1987) suggest asking the following five questions as a guide to selecting appropriate content.

1. Is the material directly applicable to the lives of our learners now?

2. Is the material consistent with the cognitive development and social maturity of the students?
3. Is the topic important in the world today and is it likely to remain important for a significant proportion of the students' adult life times?
4. Can students apply the knowledge in contexts other than science?
5. Is it a topic for which students show interest and enthusiasm?

These authors raised an important issue. How can we make room in an already overcrowded curriculum for STS education? Fortunately, a number of strategies have been suggested and tried with varying degrees of success (Jarcho, 1986).

The most popular route has been through the use of modules and/or lessons for insertion into existing, discipline-bound courses. A second could be to attack the problem by infusing STS education into existing courses through brief insertion items. A third route is to create an inter-disciplinary course in STS education. The last alternative and conceptually the most difficult one is to completely work on existing, traditional subjects by integrating throughout the course, materials on STS perspectives.

The three different conceptual frameworks given above give a good account of the understanding of STS education. A conceptual framework for teaching STS studies given by Mac Cleave-Frazier and Murray is complete in the sense it elaborates, with an example of energy. This framework is quite useful and can be practically used by educators. The second framework by R.W. Bybee gives areas, which represent the three essential themes, viz. science and technology concepts, the process of inquiry and STS interactions. It is not a very detailed framework but still a useful one to explain the concepts of STS education. The third framework – A guide to selecting appropriate content for STS course by Hickman, Patrick and Bybee – is also a useful one as it suggests ways to make room for STS education in the already overcrowded curriculum but gives only one of steps in the understanding of STS and not full account.

Curricular Framework

Different`curricular frameworks by different educators give views for curriculum construction if a separate curriculum is required for STS. What should it have? And how to go about it? Is what matters? These views will be able to give clarification to guide for curriculum construction.

A Generic Model for Curriculum Development in Constructivist Framework (Driver and Oldham, 1986)

Major phases of activity involved in planning a curriculum in a constructivist mode had been suggested by Case in 1985 and are described below:

1. Specify the values, concepts and skills which are to be the long-range goals of the educational enterprise.
2. Analyze the central structures that are employed by adults who embody these values and pass the relevant concepts or skills.
3. Analyze the developmental process of acquiring these structures by examining the problem solving of children at a variety of different developmental levels.
4. Design a set of educational activities that will be appropriate for children at each level which will also facilitate their ultimate transition to the next. In designing these activities, pay particular attention to the top level goal that is involved and make sure that they genuinely understand the rationale for any modification as elaborates which they must introduce, and that sufficient assistance and picture are available for these changes to be consolidated.
6. In implementing the curriculum, provide such additional support for teachers as is necessary to ensure that they adapt to individual differences in the children's initial level of development, in the children's preference for independent versus socially regulated problem solving, and the ease with which children can affect the sub-components of the restructuring process.

This work was further taken up by science educators, Driver and Oldham, Osborne and Wittrock, Renner, Karplus, NusBaun and Navic and Erickson. Fig. 2.3, gives six constructivist teaching sequences by (1) Driver and Oldham (1986), (2) Osborne and Wittrock (1983), (3) Renner (1982), (4) Karplus (1977), (5) NusBaun and Navic (1981), and (6) and Erickson (1979). Each gives 5, 4, 3, 3 and 3 teaching steps respectively. Though they differ somewhat from one another where they start, their overall thrust is identical. Each is designed to explicitly bring to light student's present conceptions and introduce cognitive dissonance in order that a more scientific concept can be accepted and integrated in the pupil's conceptual schemes.

They provide a good understanding of how to proceed with teaching according to different educators. This would help decide on one's steps for teaching.

This approach views the curriculum "not as a body of knowledge or skills but a programme of activities from which such knowledge as skills can possibly be acquired or constructed, though we would acknowledge that the selection of the possible learning experience is guided by the knowledge of experts" (Driver and Oldham, 1986).

Rosalind Driver and colleagues developed a model in 1986 from the children's learning in science groups at the University of Leeds and have used it for a number of years with good success. The model is an interactive one with the feedback loop resulting in continual updating of the materials in the light of increased knowledge of student's prior ideas and changing decision about teacher awareness of pupil characteristics. As given in the Fig. 2.4, there are four criteria to be decided before designing the curriculum. Firstly, the content to which the students are to be exposed. Secondly, information about students' prior ideas in the topic area. Thirdly, perspectives on the learning process and lastly, teachers' practical knowledge of students' schools and classrooms. After designing the curriculum it can be implemented and then evaluated which would again lead to modification in the four criteria described earlier. A number of lessons concerning plant nutrition, energy and the particulate theory of matter have been produced and revised using this framework as a guide and serve as useful examples for other projects to develop constructivist learning materials.

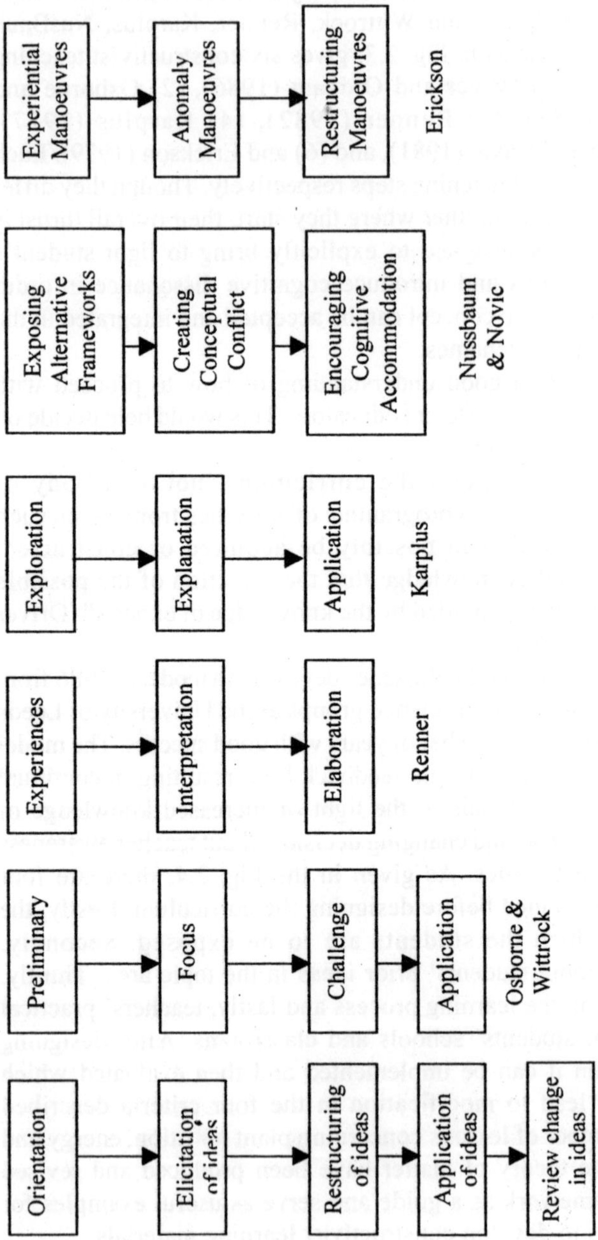

Fig. 2.3: Six Constructivist Teaching Sequences

Source: Cheek, D.W. (1992) *Constructive About Science, Technology and Society Education*. State University of New York Press.

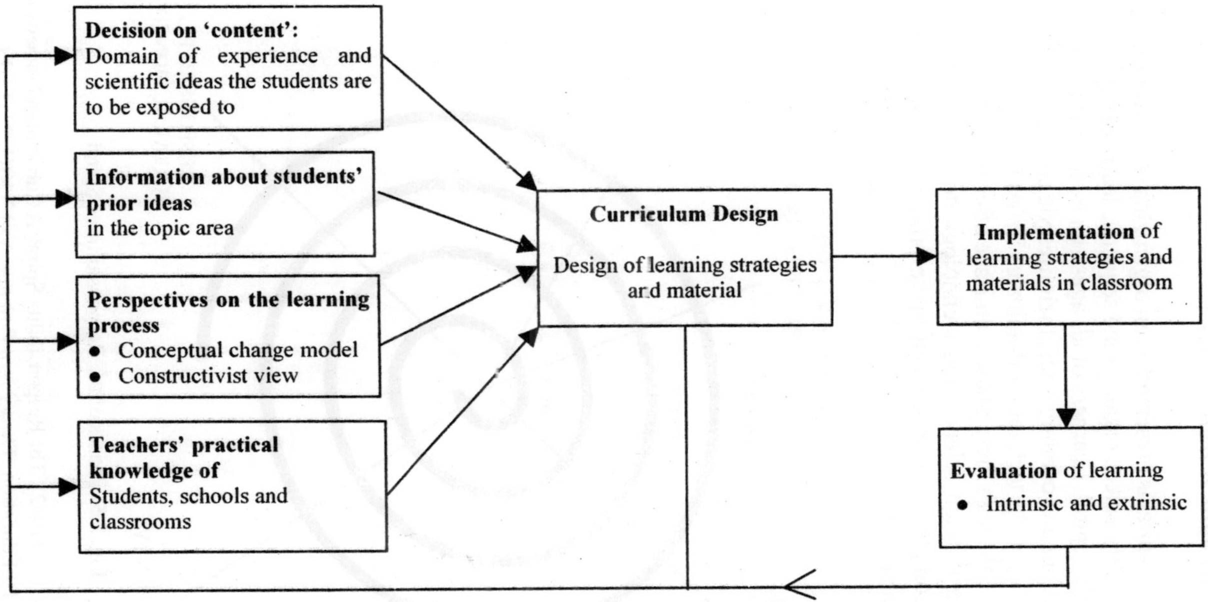

Fig. 2.4: A Generic Model for Constructivist Curriculum Development

Source: Cheek, D.W. (1992). *Constructive About Science, Technology and Society Education*. State University of New York Press.

The Responsibility Spiral: A Curriculum Framework for STS Education (Leonard J. Waks, 1992)

Responsibility spiral is presented as an organizing framework to help educators identify, select, organize and sequence learning experiences to promote this spiral. Learners of all ages can be guided in forming their connections and commitments, their lifestyle choices and values, as these bear upon the technology dominated issues facing our society. As they move through these phases, on issue after issue, confronting and thinking through science and technology dominated issues of increasing complexity, learners can make progress towards mature social responsibility.

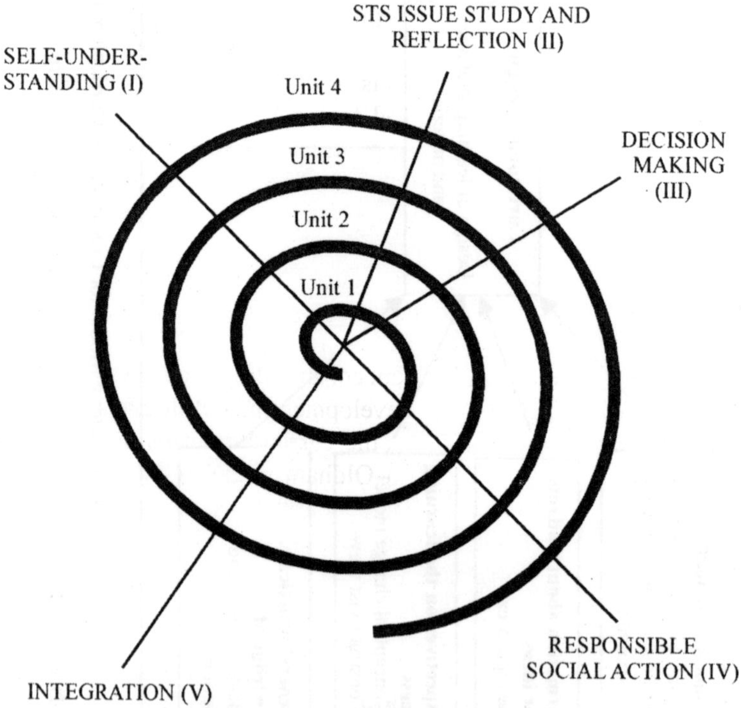

Fig. 2.5: Phases of the Responsibilities Spiral

Source: Waks, L.J. (1992). The Responsibility Spiral: A Curriculum Framework for STS Education. *Theory into Practice*, Vol.XXXI (1), 13-19.

As shown in Fig. 2.5, the spiral consists of a series of "cycle units" denoted to specific STS issues such as nuclear reactors or genetic engineering. Each unit is composed of five "phases" labelled (a) Self-understanding, (b) Study and reflection, (c) Decision making, (d) Responsible action, and (e) Integration. The Spiral, a sequence of such cycle units, is a useful organizing tool for STS units at all educational levels from childhood to adult and continuing education.

The four guidelines that follow are intended to assist teachers and curriculum developers in implementing the spiral model:

1. The STS curriculum is focused on responsibility. It begins and ends with the learner as a responsible individual and as a responsible member of society.
2. The five phases are distinguished to assure proper attention to each.
3. The STS curriculum as a whole is analyzed and, when necessary, reorganized to assure proper balance for the sequencing of the five phases of the responsibility cycle.
4. The responsibility spiral is the "inner core" of STS education. The rest of the curriculum, which may be linked to the core in creative ways, surrounds it.

A STS Curriculum Development Model (Cheek, 1992)

Based on the Driver and Oldham constructivist model, Cheek (1992) evolved an STS curriculum development model. It gives a detailed procedure for construction of the curriculum. This model mainly expands upon the Driver and Oldham model with constructivist underpinnings. Four major components contribute to the design of curriculum materials in this model: Theory (which underlines the other three), data collection, delivery system consideration and STS related content.

In Fig. 2.6, the theory component of this curriculum model entails consideration of theoretical constructs and explanatory frameworks of general cognitive psychology, the more narrow area of constructivism, moral development group processes, critical thinking and problem solving skills and persuasive communication. The curriculum team is to ascertain which areas within these research areas are pertinent to the issue under consideration.

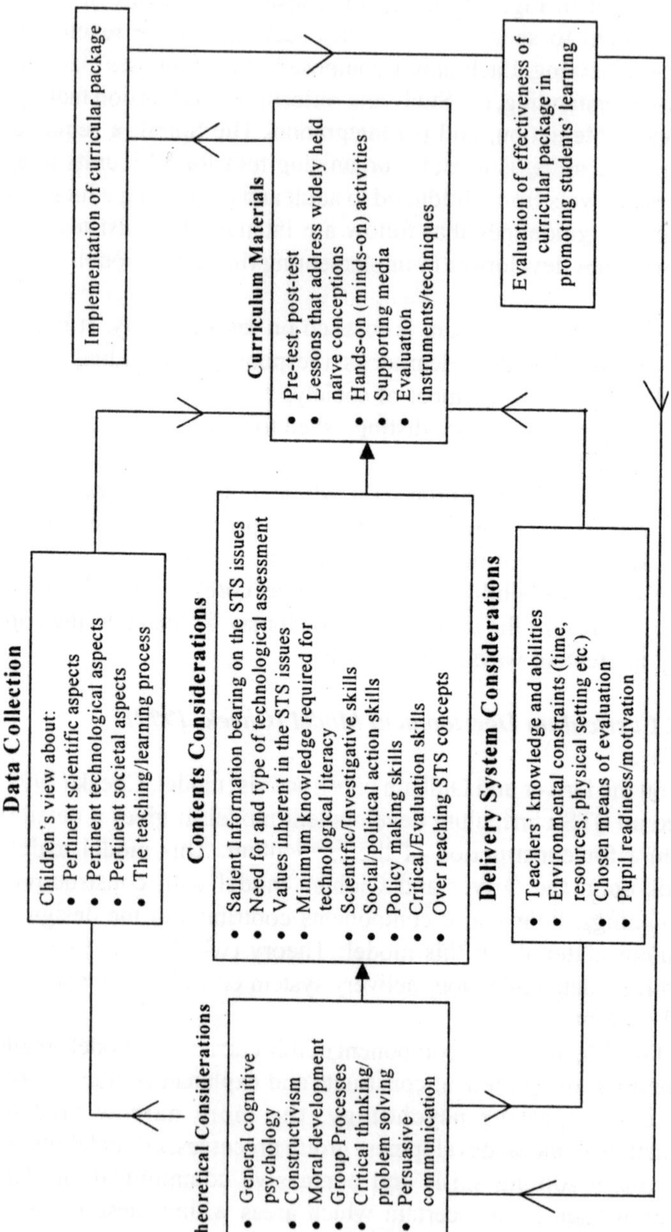

Fig. 2.6: An STS Curriculum Development Model

Source: Cheek, D.W. (1992). *Think Constructively About Science, Technology and Society Education.* State University of New York Press.

Once theoretical research has been called for salient constructs and suggested causes of action, the team can assemble appropriate instruments and techniques to collect primary data on children's views about pertinent scientific issues, technological issues, societal issues, and views of the teaching/learning process that should be consciously attended to in the design of the materials.

Theory and research should shape decisions regarding the delivery system component of curriculum materials design. Teacher's knowledge and abilities must be given priority. Materials that are beneath teacher's dignity or that stretch them beyond reasonable limits are to be avoided. The environmental constraints of schooling, such as time, resources, and physical settings must be considered. The chosen means for evaluation should be decided before curriculum materials are designed so that assessment matches design and design matches assessment. Finally, attention on the delivery system must consider pupil's readiness and motivation so that materials can be constructed that are likely to find a hearing and engage students in an effective manner.

It should also guide the content component. The features for selection of content will fall into the areas outlined. (1) Salient information bearing on the STS issues under consideration, (2) the need for and type of technological assessment regarding the selected issues, (3) the values inherent in the chosen STS issue, (4) the minimum knowledge of the issue that is considered essential for technological literacy, (5) the necessary scientific investigation skills that should be taught and/or tapped when considering the STS issue, (6) the pertinent social/political action skills involved when considering the STS issue, (7) the salient policy making skills that should be engaged when considering the STS issue, and (8) the area achieving STS concepts that are involved in a particular STS issue.

The data collection, delivery system and content components (mediated by the theory component) should then be brought simultaneously to bear upon the design of the curriculum materials. These materials should include such items as pre- and post-test lessons that address widely held naïve conceptions regarding aspects of the STS issue under review, hands on (minds on) activities, supporting media and evaluation instruments and techniques. A feedback loop to the theory component reminds that theory based considerations should be kept in mind. Moral dilemmas that are raised by aspects of STS issue should be developmentally appropriate. Theoretical considerations may suggest revision even before the material undergoes formative evaluation.

When the materials have reached their initial design plateau, the implementation of the curricular package should occur. Ideally, the formative evaluation audience should be diverse and large enough to provide sufficient data to uncover likely difficulties with the materials. Evaluation of the effectiveness of the curricular package in promoting student learning places the onus of success upon the materials, not upon the student. If a significant number of students fail to achieve a specified goal of the materials, the fault lies with the materials, the theoretical construct, improper data collection etc. but not with the students. This is the reason for the feed back loop taking us back to earlier aspects of the model. Careful attention to these steps may make it clear that the fault lies with a theory and substantial modifications to the currently held theory may be required. The initial data collection scheme might be found to be followed in some fundamental ways, resulting in an incomplete or misleading generic set of students' conceptions regarding particular aspects of an STS issue. The delivery system chosen may have been inappropriate due to faulty considerations within its boundaries. Content may have been mistakenly selected that was not age appropriate, or skills needed for a task were overlooked. In short, the entire model envisioned here is dynamic one with continual feedback and critical self-appraisal by team developers throughout the process.

Such a system approach leads to clearer conceptualization of the task of STS curriculum development and results in materials more appropriate for both teachers and students. The degree to which a team of developers could tap into these components through the process of curriculum development would directly enhance the efficacy of the final product.

Specific Curricular Projects

There are specific curricular projects which are real strategies towards building of STS instruction. Various groups of educators and associations proposed various models, projects, courses etc. out of which most were tried out and a good feedback was achieved. The author has tried to summarize these projects as per availability.

Project Leerpakpketontmikbeling Naturkunde (PLON) Project of the University of Utrecht: Netherlands (1972)

A constructivist model for STS curriculum development has been pioneered in the Netherlands for physical sciences education by the PLON (Eijkelhof & Kortland, 1987, Eijkelhof & Verbagen, 1985, Eijkelhof & Linjuse, 1988). Commenced in 1972, this Dutch project has produced over twenty modules dealing with physics and society for Dutch students from ages twelve to sixteen. The success of the units and international interest has led to the translation of some modules into English. Five modules currently available in English are titled "Ionizing radiation", "Weather changes, "Traffic and safety", "Bridges" and the simulation "Water for Tanzania". Each module is activity/experimentation oriented and raises broad physics related social issues. It took a constructivist approach to STS education in environmental issues. The PLON has identified a number of conceptions that Dutch students hold regarding various STS issues. The project attempted to design curriculum materials in such a manner as to take account of these widespread conceptions.

STS in Science: Iowa STS Dissemination Model (1982)

This model of disseminating the STS Approach has been developed at the University of Iowa. It is known as the Iowa Chautauqua Model and was incepted in 1982. It was funded by National Science Foundation (NSF, 1982) as an experimentation project proposed and coordinated by the National Science Teachers Association.

In 1984, the Iowa Utility Association expanded their programme. This provided a mechanism for supporting 30–50 teachers at each of five geographic sites in Iowa each year. With the expansion of numbers and sites, the Iowa Chautauqua Programme had then re-enrolled over 1700 K-12 teachers during the nine years of operation. The Iowa Chautauqua Model has several specific features (Blunck & Yager, 1990).

These include:

1. A two-week leadership conference for 25 of the most successful teachers from previous years who want to become a part of the instructional team for future workshops.
2. A two week summer workshop at each new site for 30 new teachers electing to try STS modules and strategies; the

workshop provides experience with STS and time to plan a five day STS unit to be used with students in the fall. Use of five day STS unit in the classrooms of all summer participants during September or early October; 3 to 5-day fall short course for 30–50 teachers (including the 30 enrolled during summer), the focus was upon developing a month-long STS module and an extensive assessment plan.

3. Interim communication with central staff, head teachers and fellow participants, including a newsletter, special memoranda, monthly telephone contacts and school/ classroom visits.
4. A 2 to 5-day spring short course for the same 30–50 teachers who participated in fall; this session focused upon reports by participants on their STS experience and the results of the assessment programme.

The work with the teachers was intended to help them teach towards achieving five major goals. These, characterized as "domains", included: Knowledge, Processes, Applications, Creativity and Attitude. Applications are goals uniquely represented by STS.

Knowledge Domain

This focuses on the more conventional content of science courses. It includes facts, information, concepts, laws, principles, existing explanations and theories being used by scientists. It was assessed by multiple choice tests with test retest reliabilities ranging from 0.76 to 0.81. It has multiple choice type questions.

Process Domain

This domain was reintroduced to science education by curriculum reforms of the 1960s. Process of science identified by the American Association for the Advancement of Science (AAAS) in the *Development of Science—A process Approach* (1963) represented a generally accepted set of processes scientists use. These include: observing, using space/time relationships, communicating, classifying, grouping and organizing, using numbers, quantifying, measuring, inferring, predicting, controlling and identifying variables interpreting data, formulating, hypothesis, defining operationally, and experimenting. Encouraging students to construct meaning with respect to processes as well as concepts is central to good instruction, including the Iowa NSTA STS approach. Current studies in Iowa are utilizing a more modern perspective concerning science processes.

Applications Domain

Science should be such that it could be useful in the everyday life of a student. Students need to become sensitized to those experiences they encounter that reflect ideas they have learned in school science. Some dimensions of this domain are: seeing instances of scientific concepts in everyday life experiences; applying concepts and skills to everyday problems; understanding scientific and technological principles involved in household services; using scientific processes in solving problems that occur in everyday life; understanding and evaluating mass media reports of scientific developments; making decisions related to personal health, nutrition, and lifestyle based on knowledge of scientific concepts rather than on "hearsy" or emotions; integrating science with other subjects; specific actions designed to realize problems and/or contribute to the improvement of local, regional, national and/or international problems; becoming involved in community-action projects; extending school experiences beyond the classroom; and emphasizing the inter-relationships and interconnectedness of science to other human enterprises.

Creativity Domain

Some of the human abilities important in this domain include: visualizing and producing mental images; combining objects and ideas in new ways; producing alternative as unusual uses for objects; solving problems and puzzles; predicting possible consequences suggesting possible causes; designing devices and machines; producing unusual ideas, identifying; isolating, merging; diverging; and converging. Creativity is defined for the Iowa STS studies to include both the quantity and quality, and uniqueness of questions, possible causes, or predicted consequences of given situations. STS instruction often focuses on questions, causes for objects and events and consequences for certain action or happenings (Yager, 1989).

The creativity domain is based on visualizing and imaging, combining ideas and objects in new ways and identifying unique uses for certain objects. The average test retest reliability for test compiled by the teachers was 0.88.

Attitude Domain

This domain includes: developing positive attitudes towards science in general, science in school and science teachers; developing positive attitude towards oneself; exploring human emotions; developing sensitivity to and respect for the feelings of other people, expressing personal feelings in a constructive way; making decisions about social environmental issues and exploring arguments on various sides of an issue. The focus of STS upon current issues and student questions and interests make the attitude domain particularly important.

The Attitude Domain was assessed by a collection of items from the National Assessment of Education Progress (NAEP, 1978). An 18 Likert type items five-point scale was used (Yager *et al.,* 1990). Focus on the acquisition of knowledge first may be a major problem with most science teaching. It may be faulty logic to argue that students first must be given knowledge before they can hope to apply it. Perhaps instead, basic science knowledge can be seen as useful, as questions arise and so more experience accrues as students work in the application domain. Fig. 2.7 is an attempt to illustrate the Iowa plan for STS instruction and evaluation.

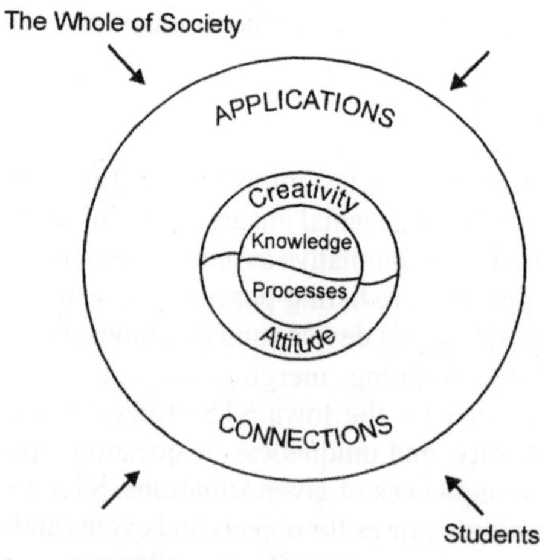

Fig. 2.7: The Iowa STS Project: Student Growth in a Variety of Domains

Source: Yager, E.R. & Brunkhorst, J.B. (1984). *Science Education Centre*, the University of Iowa, Iowa City, USA, p. 646.

The results of the Iowa model show that knowledge areas have rarely shown any difference in the acquisition of the knowledge whether students experienced STS content or traditional content. Affective domain results have been dramatic. Students experiencing science in STS format have more positive attitude. They have more realistic views concerning the utility of science study and these attitudes are maintained across grades. Since most typical science classes never considered technology (application of science) or attempted to connect the knowledge/process exposure/acquisition to real life situations, STS students perform better than students who experienced traditional science.

Science in a Social Context (SISCON), Addinell and Solomon (1983)

One of the most widely known secondary STS programmes has been Science in a Social Context (SISCON) from Great Britan. One of the first school STS courses, SISCON-in-Schools, chose this simple way of breaking into schools and their examinations structure, in 1981. The original SISCON (Science in a Social Context) project was started in 1970 to promote STS teaching in universities and polytechnics. It produced teaching materials which linked science with economics in both the developed and developing countries, and others which taught the new studies of philosophy and sociology of science, or studies of the connection between science and warfare. However, while it still remained confined to degree courses, at least one strand of STS—the intention of bringing science to a wider audience so that the public could be empowered to take part in public and political issues—could not be realized. For that it needed to go into mainstream schools. In 1978, SISCON-in-Schools was begun by the few school teachers who had attended the parent organization's summer schools. Like the earlier Science in Society this scheme was for sixth-formers, but it differed from that course in the strands of STS which SISCON chose to take up. Where the earlier course was strong on economics and industry, SISCON focused more on the human and fallible nature of scientific theories, quality of life in the first and third worlds and the social effects of new technologies including armaments. The SISCON materials provide a good role model for other developers to emulate—particularly in nations whose curricula are according to set national syllabi (Lewis, 1987).

Science through Science, Technology and Society Project (1985)

Funded by the National Science Foundation from 1985-88, the project developed the following eight criterion as a set of guidelines for instructional materials evaluation in STS:

1. The relations of technological or scientific developments to societal relevant issues are made clear early and in compelling ways to capture attention.
2. The mutual influences of technology, science and society on each other are clearly presented.
3. The material develops learners' understanding of themselves as interdependent members of society and society as a responsible agent within the ecosystem of nature.
4. The material presents a balance of differing viewpoints about the issues and options without necessarily striving to hide the author's perspective.
5. The material helps learners to venture beyond the specific subject matter to broader considerations of science, technology and society, which includes a treatment of personal and societal values/ethics.
6. The material engages students in developing problem-solving and decision-making skills.
7. The material encourages learners to become involved in a societal or personal course of action after weighing the tradeoffs among values and effects drawn from various scenarios or alternative options.

The material uses this STS linkage to foster learners' confidence in handling and understanding at least one limited science area, and/or handling and using some qualifications as a basis for judgments in the STS area.

Project on Topics to be Covered in STS Courses, Bybee (1987)

The importance of basic science knowledge for an adequate understanding of STS issues is obvious and always finds its way into recommendations. Less frequently acknowledged on the part of science educators is the need for basic knowledge in the areas of technology and society.

What kind of topics should be covered in an STS course in science education or selected for insertion in an existing science course? Rodger Bybee and his associates explored this question with 262 science educators from forty-one countries in the spring of 1984. The top twelve issues that emerged from this survey, in order of perceived importance, were: (1) World hunger and food resources, (2) Population growth, (3) Air quality and atmosphere, (4) Water resources, (5) War technology, (6) Human health and disease, (7) Energy shortages, (8) Land use, (9) Hazardous substances, (10) Nuclear reactions, (11) Extinction of Plants and Animals, and (12) Mineral resources.

The science educators suggested the minimum amount of instructional time devoted to science related social issues in percentages, i.e. for elementary school 10 per cent, middle/junior high school 15 per cent, high school 20 per cent and college/university 25 per cent. According to them, these percentages leave ample time for development of other science educational goals such as knowledge, inquiry and career awareness. Rather than diluting these goals, there is every possibility of students learning more about science and technology because it will be presented and experienced in a context with personal and social meaning.

Topics in Applied Science: A Model STS Curriculum (Carole P. Mitchener and Ronald D. Anderson, 1989)

A model STS curriculum was cited as an exemplary Science Programme in 1982 by the National Science Teachers Association. This curriculum was developed and implemented in Jefferson County R.I. School District, a suburban area of Denver, Colorado. With a core visiting team of 12 Jefferson County School district teachers and co-ordinators, topics were developed during the late 1970s through a series of summer workshops and in-service sessions. It continues to be revised through the same vehicles.

Topics step well beyond the confines of a traditional science course. It is consistent with the recommended science education guidelines spelt out by that endeavour. In addition, it is based on the use of the co-operative learning approach.

The content of topics in applied science is from the earth science and physical science disciplines with a Science/Technology/Society focus. The purpose of the course is to promote scientific literacy in junior high students. It provides students with the opportunity to apply scientific and technological knowledge in making decisions about real community situations.

Topics are very consistent with current developments and trends in science education. The course also employs cooperative learning (Johnson, 1975) as the instructional strategy over 50 per cent of the time. While topics are designed for all students, it is especially designed to provide success for students with average or below average interest in science. It is an alternative science course for students who have no post-secondary plans in science.

Topic is divided into four nine-week self-contained units. The themes for the four units are technology, survival, energy and land use. Each unit is then divided into sections which are centered on one question and contain a series of sequential and activity-oriented lessons. For example, the unit entitled "what technology and you is about" is divided into four sections: How do I use it? How do I choose it? How is it created? And how is it changing my world?

The British Columbia STII Course (Uri Zoller et al., 1990)

Based on the belief, shared by science educators and Ministry of Education in British Columbia (B.C.), that students should have a better understanding of technology and science for their interaction with society, the course Science and Technology (ST) was proposed in 1984 for students in grade II, not electing one of the existing courses.

Based on the commonly accepted philosophy and rationale for S/T/S courses (Aikenhead, 1985; Bell, Watts and Ellington, 1985; Bybee, 1985; and Zoller, 1987) the following three super-ordinate goals have been established for the STII course by the curriculum committee of B.C.

To provide students with opportunities to:

1. Develop an appreciation of the interactive nature of science, technology and society;
2. Gain knowledge of technologies as applications of science; and
3. Develop the ability to respond critically to technological issues.

In an attempt to achieve the above goals the STII course took an approach different from the traditional one to science teaching.

The STII course uses the S/T/S orientation as its conceptual base, focusing on the exploration of contemporary controversial, significant, technological issues. At the heart of the course is the student as an involved citizen understanding and reacting to issues that have no absolutely right and wrong answers (Science and Technology 11, 1986).

There is no textbook for the course, the course consisted of 16 modules (covering 16 different topics) divided into three main categories: Personal connections, Local/vocational connections and Global connections. Five of the modules—introductory module, health technologies, and the computer in the work place, telecommunications and the future—are mandatory core modules. The remaining are "optional" modules, four of which are to be chosen by each class teacher in preferences; Class discussions/debates and opinion/views presentation by the students concerning S/T/S issues are major teaching/learning strategies in the course (Curriculum Development Branch, 1986).

The SATIS Project, Patric Fullick (1992)

The Science and Technology in Society (SATIS) project was set up by the Association for Science Education (ASE) the professional organization of UK science teachers, to produce materials that would show young people in the 14-16 age group how science and society are inter-dependent (Holman, 1985b).

Writing of units was carried out by volunteer teachers and other science educators, with advice on specific aspects of units being given by experts in the particular field concerned. All units were extensively tested in a range of schools, the feedback from these being incorporated into the final version of the units. A typical SATIS unit contains some reading (to convey new ideas and to reinforce previous work), questions, and an activity of some sort. One hundred units have been produced, and copies have been sold to nearly every school in England and Wales. Topics covered in SATIS units include homeopathic medicine, safety in road transport, invitro fertilization, the development of nuclear fusion as a power source and several others.

In preparing SATIS materials, it was considered necessary to keep in mind the demanding criterion that such materials must satisfy if they are to be widely used. These criteria include:

Cheapness—Teachers will not consider material if it is expensive to acquire and/or use. This ruled out the production of a textbook as a resource.

Ease of Use—If it is to be widely used, the material must not need large amounts of teacher's preparation time.

Relevance—Not only does material need to be relevant to young people's lives, it needs to be relevant to the syllabus their teacher is following. SATIS units were developed using the "Science first" model instead of "Application first". The final format of resources was agreed to be as follows: (i) A bank of resource units (not a new course), (ii) produced as reproducible masters to be copied by schools. Units were to be active and varied in approach, including strategies like discussions, home surveys, data analysis, case studies, simulations and role played in decision-making exercises and experimental work.

The project has twice undergone extension, once in January 1991 and later in 1993, seeking to encourage young people to think about and exchange their perceptions of the science society interface in a European context.

The evaluation exercise was a valuable experience for all the members of the project team and confirmed the initial ideas that material must be relevant in two ways if it is to be well used by teachers and liked by students. Those two forms of relevance are:

- Relevance to syllabus.
- Relevance to students' lives either as a starting point or as something that will clearly be of use in the future.

The SATIS evaluation has proved that young people enjoy learning about science when they can relate it to social issues and applications, and that they are enthusiastic about learning in this way (as are their teachers!), but much more work is still to be done before we can predict this with any confidence that our population of the future is overwhelmingly scientists.

The Integration of STS into Science Education (Glen Aikenhead, 1992)

Aikenhead describes the major features and characteristics that a new generation of science text books require.

STS science materials are best organized in a manner depicted by the arrow in Fig. 2.8. The arrow maps out a preferred sequence of teaching events (Eijikelhof and Kortland, 1987), which are summarized in this section. STS instruction begins in the realm of society, represented by the box that forms the outer boundary of the figure. A key question or problem is posed, for instance, should we be concerned about high voltage power lines in our community? Or, how can we explain the conflicting scientific testimony in a newspaper article?

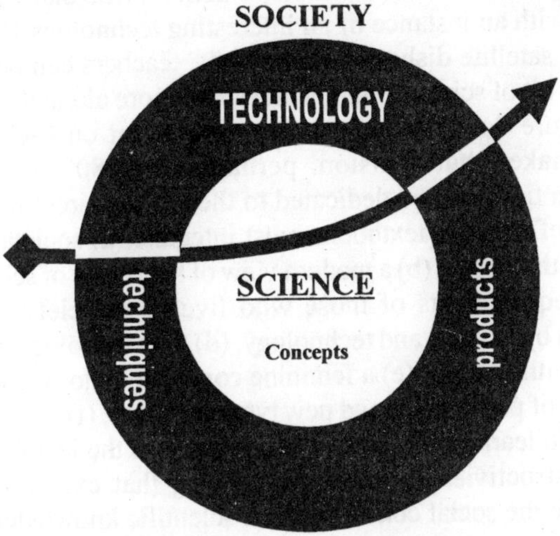

Fig. 2.8: A Schematic Illustration of a Possible Sequence for Organizing STS Material (modified Eijkelhof & Kortland, 1987).

Next, there is usually some technologies with which students need to become familiar with even at a superficial level in order to understand the society-based question or problem. The black doughnut in Fig. 2.8 represents the domain of technology.

A social question or problem (the beginning of the arrow in Fig. 2.8) creates the need to know certain technological knowledge (the doughnut area in Fig. 2.8), but both create the need to know some science content. This science content helps students understand the technology well enough to make thoughtful decisions on a related social issue (the arrowhead in Fig. 2.8).

In STS science, traditional science content got watered down (Ziman, 1980) but it is embedded in a social technological context (described by Fig. 2.8). The choice of content is made on the basis of its meaningfulness to students and the science content was logically generated on a need to know basis required for a particular science curriculum.

The sequence of events suggested by the arrow in Fig. 2.8 moves from the original domain of society, through the domains of technology and traditional science and then out again into technology and society. Finally, the arrow in Fig. 2.8 ends in the domain of society. Fig. 2.8 suggests a sequence that is useful for a lesson, a unit or for a textbook in STS science. Variation in lessons or units can be achieved by starting a class in the technology domain (the black doughnut in Fig. 2.8) with an instance of an interesting technology (for example, television satellite dishes). Alternatively, teachers can begin with an intriguing bit of science content and some more along the arrow from there. Figure does not indicate the time spent on each domain. A teacher makes that decision. perhaps 60 to 80 per cent of the instruction time will be dedicated to the central circle in Fig. 2.8.

In brief the STS textbooks must integrate (a) societal issues and interact with science, (b) a modern view of the nature of science, (c) the literacy requirements of those who live in a society increasingly dominated by science and technology, (d) the technological world that interacts with science, (e) a feminine contribution to science (in terms of number of participants and new types of ideas), (f) a student centred approach to learning from the topics chosen to the learning used, and (g) a constructivist approach to learning that exemplifies, where appropriate the social construction of scientific knowledge itself.

A Comparison of Guidelines and Curricular Framework for STS Education

Dennis W. Cheek (1992) in his book *Think Constructively About Science, Technology and Society Education* selected fifteen researchable comprehensive documents that attempt to convey the nature of STS education. In specific, educational goal statements were selected for comparison. The first statement by Peter Rubba and Randall Weisenmayer (1985, 1988) adapts the four-tired goal statements for environmental education of Hungerford, Reyton and Wilke (1980) to STS education.

The second individual statement is that of Leonard Waks (1987) who chaired the programme committees of the Second and Third Technological Literacy Conferences and acted as editor or co-editor of the Proceedings of the Second, Third and Fourth Technological Literacy Conferences. Waks is one of the prime theorists within the STS community at large and in touch with a wide range of viewpoints. His academic preparation as a philosopher of education and organizational psychologist contributes to his acumen in framing a technological credo that would likely find a fair degree of consensus within the STS community.

A third individual statement is that of Glen Aikenhead (1986), a prominent Canadian science educator who has been active in the STS field for some time serving on the board of directors of the National Association for Science, Technology and Society, and as secretary of the STS Research Network for the International Organization for Science and Technology Education (IOSTE).

Statements from organizations and curriculum projects selected for analysis were those of the Science through Science, Technology and Society (S-STS) Project (1985); the STS syllabus from the British General Certificate of Secondary Education (GCSE) of the Northern Examining Association (1988); the Science and Technology 11 (ST11) Project from the Ministry of Education, Province of British Columbia (1986); the National Science Teachers Associations (NSTA) guidelines for STS education (1982); the Social Sciences Education Consortium's (SSEC) curriculum framework for STS (Hickman, Patrick and Bybee, 1987); Science in Social Context (SISCON) from Great Britain (Addinell and Solomon, 1983); the Biological Sciences Curriculum Study's (BSCS) programme "Innovations" (Biological Sciences Curriculum Study, 1984); the position paper and guidelines for STS education from the National Council for the Social Studies (1983); Block J (BJ) of the middle level science syllabus from the New York State Education Department (1985); the Project Synthesis (PS) report from the committee dealing with STS (Piel, 1981); "Preparing for Tomorrow's World" (PTW), originally prepared for environmental education but marketed to STS educators (Iozzi, 1987); and finally, the new position paper and guidelines for STS education from the National Council for the Social Studies (1990).

Cheek and Foltz (1987), identified an initial set of eighteen general descriptors as comprehensively covering areas mentioned in a smaller set of twelve documents for a U.S. Department of Education study of STS education for urban and minority youth. Additional eleven descriptors were added from a set of fifteen documents. These descriptors are viewed as important areas of concern for STS education that have been raised in the documents. Their appropriateness or relative importance in regards to STS education is not addressed here.

The twenty-nine descriptors were grouped into four general areas: knowledge, skills, ethics and values, and action (in the sense of local and/or national involvement in the community of society) as given in Appendix A.

The knowledge area covers fourteen aspects, i.e. (i) describes interaction of science, technology and society, (ii) understand limits of science and technology, (iii) recognize and/or distinguish scientific and technological developments, (iv) awareness of current STS issues, (v) awareness of global inter-dependency, (vi) understand key STS concepts, (vii) distinguish personal opinion from scientific evidence, (viii) understand application of technology, (ix) be able to use correctly some qualification techniques, (x) critique proposed solutions to STS issues, (xi) know that STS issues are persistent, (xii) understand some of sociology and epistemology of science and technology, (xiii) nature of inquiry and (xiv) science and technology are controlled by society.

The skill areas cover eight aspects, i.e. (i) communication skills regarding STS issues, (ii) science or cognitive process skills, (iii) decision making and problem solving skills, (iv) critical thinking skills, (v) group process skills, (vi) group leadership, (vii) conflict management and negotiation, and (viii) use technology in practical situations.

Ethics and values consist of five aspects, i.e. (i) aesthetic appropriation of world, science and technology, (ii) value democracy and democratic process, (iii) understand importance of values, (iv) be able to articulate personal values, and (v) identify values bearing on any given STS issue.

The action area consists of two aspects i.e., (i) conduct research and (ii) take action on STS issue in local community and/or society at large.

Each document was read to determine which of the twenty-nine general descriptors were mentioned in some ways within the document. Some latitude was permitted in the exact phrasing of the goal statement and the document did not have to mention all aspects in the descriptor. For example, Knowledge Descriptor 2 says: "Understand limits of science and technology". If the document simply mentioned science and not technology or *vice versa* it was still considered present. In short, the resulting matrix was overly generous rather than conservative.

As described above, there is a widespread over the twenty-nine selected descriptors. The most recent documents are the most comprehensive in terms of these descriptors. Only one descriptor ("Describe interaction of science, technology and society") found unanimous agreement in all fifteen documents. Only eight out of twenty-nine descriptors found agreement in greater than fifty per cent of the documents. The description probably does not directly reflect the total viewpoints of the organizations or individuals represented regarding STS educational goals. Individuals and organizations may well subscribe to more goals than articulated within their respective documents. Some differences may be due to grade levels targeted. Block J from the New York State Education Department is part of a science syllabus for Middle and Junior High Schools. Some goals for senior high schools may not have been deemed appropriate for this level and are not mentioned. A new middle level science syllabus under development in New York State already reflects a broader view of STS. The recently adopted NSTA position paper on STS education also reflects a broader perspective than the 1982 document used here (National Science Teachers Association, 1990).

These results suggest that serious attention be given by the STS education community to formulating a comprehensive set of statements that might find wide consensus within the STS community.

A serious shortcoming of nearly all these documents is the lack of attention to a substantial body of literature in the areas of developmental and cognitive psychology, group social dynamics and selected domains within social psychology. Only two documents explicitly mention psychological underpinnings for curriculum development and instructional efforts in STS education (Iozzi, 1982; National Council for the Social Studies, 1990a).

The discussions on the conceptual and curricula frameworks of STS clearly portray that as an organised discipline it is still in a formative stage. How it will evolve is yet to be determined. STS instruction needs to be defined by local educators and the content and instructional approaches are to be thoughtfully developed with equal considerations given to the various demands placed upon schools and with the important task of educating students to live in an increasingly technological world. There is serious need to consider the result and implications of these overlooked research areas to appropriately tailor our goals and curriculum development effort to the needs and abilities of students in elementary and secondary schools. The infusion of STS theme into the already existing content offers unique opportunities to achieve these goals.

3

Review of Related Literature

Introduction

This chapter deals with the analysis of the related literature which includes dissertation abstracts, research reviews, research journals, works and seminar papers. As STS is an upcoming topic of research, it is important to review it both theoretically as well as empirically, thus both theoretical and empirical studies are included in the review.

The studies related to these aspects are resumed separately and presented under different sections of the present chapter. They are divided into various sub-sections according to the categorization of studies having similar objectives or emphasis.

The studies were grouped into the following categories:

(i) Studies related to the need for Science, Technology and Society (STS) curriculum.
(ii) Studies related to evaluation of science, technology and society content in the textbooks in terms of STS.
(iii) Studies related to students', teachers' and science educators' beliefs and perceptions about STS.
(iv) Studies related to the effectiveness of STS.

Studies Related to the Need for Science, Technology and Society (STS) Curriculum

Reform is needed in both the content and methods of instruction of science education. The deficiencies in the areas of goals, curriculum and instruction are the reasons why a majority of students believe that topics they study in science classes are dull and boring, are 'no fun' and have little relevance to the real world. In the 1960s, 'a knowledge explosion in science' and developments in learning theory combined to provide an overwhelming stimulus for change. New curricula were developed for use by classroom teachers. Despite the best intentions there was not enough evidence from science education status studies of the 1970s that classroom practice actually changed as a result of exposure to quality curriculum materials. The STS movement has grown as a result of a general feeling of dissatisfaction from the outcomes of the curriculum reforms. STS was introduced in the educational system to bring about meaning to education and to check the drawbacks of previous curricular reforms. This section of the review deals with studies, which portray that there is a need to move towards a STS curriculum.

Hurd (1981) presented the findings of the group looking at data from biology education, which emphasized the significant discrepancy between biology curriculum in schools, and the changing, technologically oriented American society. Emphasis in biology classes was on rote memorization of information. There was little attempt to relate biology education to everyday life or current biologically related social issues. The use of inquiry methods in teaching was notably absent in American biology classrooms despite the fact that biology lends itself particularly well to the use of such methods.

Hurd (1985) has described an acceptable science curriculum for pre-college levels that he believes have cultural as well as scientific and technological validity. He identifies the following elements as essential for the reformation of science education: (i) yearly required instruction in science for all K-10 students, (ii) organization of courses in a social context rather than in special disciplines; (iii) a balance of science and technology with an emphasis on inter-relationships between and among them, values, and society at large; (iv) concentration on responsible decision-making and critical-thinking skills; and (v) development of courses around persistent social problems that have underlying health, environmental and technological aspects.

Penick and Yager (1986) in the article "Science Education, New Concerns and Issues" describes the crisis in science education in the United States. Many organizations and groups took up further research and synthesis efforts. One of the most comprehensive of these synthesis efforts, Project Synthesis, involved 23 leading researchers in science education. The project identified four goal clusters—four reasons for inclusion of science in every school programme: (i) personal need—science education should prepare individuals to utilize science for improving their own lives and for coping with an increasingly technological world. (ii) societal issues—science education should produce informed citizens prepared to deal responsibly with science related societal issues (iii) career education/awareness—science education should give all students an awareness of the nature and scope of a wide variety of science and technology related careers open to students of varying aptitudes and interests. (iv) academic preparation—science education should allow students who are likely to pursue science academically as well as professionally to acquire the academic knowledge appropriate for their needs. The listing of crisis condition in science education reveals that **teacher education must respond to the real crisis with proposals designed to offer corrections.** It is time to reflect on what schools and teachers are trying to do, to reassess goals, and then measure success in terms of such goals. A good science education for the public is basic to good citizenship and a good teacher education programme is its prerequisite. Ours is a scientific/technological society and all persons need a better, more realistic and more useful preparation for it.

Baez (1986) in the article "Science Technology and Society: Educational Implications" concentrated on one science, technology and society topic, namely the environment. According to him, the environment is a theme which represents an overriding concern and which can also lend unity to treatment of the subject of STS. The author leans on related issues of environment and conservation of resources as a means of illustrating how innovation in science and technology education can be more closely linked with the needs of society. The writer concludes by saying that this approach will be different from the usual approach to the teaching of science and technology in the sense that the behavioural changes which are aimed for are considered just as important as the scientific information which is learned by the technological skills which are acquired. He suggests not merely the creation of new environmental education courses but also the development of instructional methods and materials which can be integrated into existing courses and other types of science and technology activities, both in and out of the class room.

Bybee (1987) in his essay 'Science Education and Science–Technology–Society (STS) Theme' aimed to clarify the relationship between science education and STS theme and to develop justification for the position that the STS theme is rightfully a part of our educational purpose. The position is developed by first examining the current debate over the definition of science education's purpose as a social institution. Finally, there is discussion of STS in science education, some issues engaged by including this theme in science education programmes and practices. The conceptual framework, which the author proposed, included primary areas in which objectives are formulated, i.e. scientific and technological knowledge, process and personal/social applications. Three essential themes, namely science and technology concepts, the process of inquiry and science-technology-society interactions were included in the conceptual framework. Overall review of the conceptual framework suggests an orientation towards the personal and social purposes of science education while examination of the separate columns identifies specific objectives in the area of knowledge skills in S-T-S interactions.

Khan (1987) studied the status of science and technology in Botswana. A postal questionnaire consisting of fifty items was given to thirty-two trainee students of diploma and degree (science teachers) at the university and sixty-two professionals (technical and managerial) and politicians. Questions attempted to determine some general knowledge of science, attitude to the type of science appropriate to Botswana, the role of government in science policy, and environmental awareness. The questionnaire was in two main sections. In the first section, respondents had to select the most appropriate of four answers to a question/statement. In the second section, respondents were asked to rate items in importance on a 0 to 4 scale. The findings revealed that science education in Botswana is not up to the mark. With this as well as the need to establish some baseline evidence of how science and society issues are perceived in mind, a pilot survey was conducted. A questionnaire was given to science teachers and professionals. The findings suggested that: (i) Much more emphasis needs to be given to developing conscientiousness in a university students towards environmental issues. (ii) Students did not feel as strongly about these issues as those in the world of work. (iii) The educational elite had a fairly sound perception of science and technology and this is shared by the next generation of science teachers. There were some indications that educational level does influence factual knowledge of general science issues.

Okebukola and Jegede (1990) studied eco-cultural influences upon students' concept attainment in science. The prediction that eco-cultural variable could exert influence on students concept attainment in science was tested in this study using a 2 (general environment) X 2 (reasoning pattern) X 2 (nature of home) X 2 (goal structure) fixed–effect ANOVA design. The results showed that (1) students who live in a predominantly automated environment did better than those in a predominantly manual environment; (2) students whose reasoning patterns were predominantly magical and superstitious performed significantly lower than those who were empirical in reasoning: (3) rural dwellers were predominantly co-operative in outlook; (4) students from authoritarian homes achieved less on the science concept test when compared with those from permissive homes.

Hart and Robottom (1990) in the article 'The S-T-S movement in science education: a critique of the uniform process' critically examines the programme and professional development process that has characterized past and current science curriculum reform effort in an attempt to illustrate fundamental educational reform problems. The need for new directions in science education is uncontested here. The issue is the problematic nature of the reform process itself. The article centered on the fact that the new methods of reforms must be researched, methods that directly involved practitioners in critical reflection on science curriculum development.

McFadden (1991) in the article "Towards an STS School Curriculum" gives the experience of a Canadian Science-Technology-Society (STS) curriculum project. It is presented in this article as an indication of the problems that are likely to follow from attempts to introduce an STS curriculum piecemeal. These problems include redundancy within school curriculum, and reduced time for science concept and skill development. These problems are likely to undermine the credibility and long term acceptability of the STS proposals. It is argued here that only a coordinated reform of the entire school curriculum can have the result that all the science related aspects of curriculum are addressed without loss of attention to science concepts and skills development.

Bingle and Gaskell (1994) explored scientific literacy for decision making and the social construction of scientific knowledge. They explored the implications, for this process, of two broad different points of view, which they term as the positivist and social-constructivist positions. Finally, they discussed the implications of this analysis for teaching about science, technology, society (STS). To better illustrate their argument, a case involving a current science related issue was used—NASA's Galileo Mission to Jupiter. They concluded that the adoption of social constructivist view of scientific knowledge and its inherent way of evaluating knowledge claims clearly has implications for future approaches to STS education. However, this way of teaching STS may be difficult to implement. This may be because the current social issue approach is most compatible with traditional science content as it is now taught. The shift to the approach suggested above would require a more drastic reorganization of the curriculum, one that may be resisted by the current stakeholders in science education.

Hansen and Oslon (1996) studied how teachers construe curriculum integration: the science, technology, society (STS) movement as *Bildung*. The concept of *Bildung* has the literal meaning of shaping an object of a behaviour—a kind of making. It became a pedagogical concept during the German enlightenment between 1770 and 1830 when *Bildung* was construed as liberal education for everyone. Today the crisis of modernity has contributed to a revival of the ideal of *Bildung*; a moralty-based science education is needed. A better science education can be created based on the notion of *Bildung* which has a reflective dimension asking the individual to determine his or her position in the surrounding world. It deals with the change in teacher thinking as it is necessary to cope with integrated science and STS education. It involves reflection on the moral basis of the subject matter and its teaching. Discussions were held with five teachers and teacher educators at Queen's University, Canada on issues in technology education and STS. Discussion was on shift of curriculum emphasis towards integration. Data was also gathered from teacher interviews in a German basic integrated science education project. Based on interviews with Canadian and German teachers the authors noted that while teacher thinking is rooted in the discipline, and the discipline is the context for teacher reflection about STS approaches, it is also a source of constraint. This constraint raises consensus about the possibility of STS science. The authors consider how integration might enable teachers to overcome the elementrization of school knowledge and work towards helping students cope with the problems of real life. Curriculum material has to demonstrate how inter-disciplinary themes support the new approach. It is not sufficient to make science lessons more relevant to the students through technical applications or inter-disciplinary knowledge organization. Teachers need new role models, practical experience and good teaching material. A discourse has to be established that replaces the virtue of elementarization by a virtue of continuous reconstruction of science content for practical action. Integration transforms the traditional virtues of the teaching profession from talking about well-defined facts to the ambiguities inherent in social and technical issues. Values become integrated into science education instead of remaining with subjects like religion or language. The new thinking is based, among other things, on a non-destructive man-nature relationship.

Pedretti (1996) in the article "Learning about science, technology and society (STS) through an action research project: Co-constructing an issue based model for STS education" describes the experiences of six science teachers and a facilitator (the author) involved in an action research group in science, technology and society (STS) education. Two strands of researches have influenced the evolution of this action research project. The first strand comes from an interest in science, technology and society education, while the second strand emerges from the literature on action research as a possible means of accomplishing some of the goals articulated by STS advocates. Therefore, the purpose of the paper was two-fold. First, the paper explored the genesis of an issues based model for STS education and curriculum development created by the teachers and the facilitator of the action research group and briefly highlighted applications of this model in a classroom context. Second, action research was examined as a potential tool for professional development in the context of science education.

Aikenhead (1997) in the article "Towards a first nations cross-cultural science and technology curriculum" explores First Nation (Native American) science education from a cultural perspective.

Science is recognized as a sub culture of western culture. Scientific and aboriginal ideas about nature are contrasted. Learning science is viewed as culture acquisition that requires First Nation's students to cross a cultural border, from their everyday world into the sub culture of science. The pathway towards the cross-cultural education explored in the article was: (i) founded on empirical studies in educational anthropology, (ii) directed by goals of First Nations people themselves, (iii) illuminated by a reconceptualization of science teaching as cultural transmission, (iv) guided by a cross-cultural STS science and technology curriculum, and (v) grounded in various types of content knowledge (common sense, technology and science) for the purpose of practical action such as economic development, environmental responsibility and cultural survival. Cross-cultural instruction requires teachers to identify cultural border crossings for students and to facilitate those border crossings by playing the role of tour guide, travel agent or cultural broker while sustaining the validity of students' own culturally constructed ways of knowing.

> Thus the studies showed that emphasis has been on the rote memorization of information. There has been very little attempt to relate education to everyday life related social issues. There is a need to reassess goals of science education and develop instructional methods and materials which can be integrated into existing courses. Organization of courses should be in social context rather than in special disciplines. STS themes should be included in science education and action research should be used as a professional tool for development in science education.

Studies Related to Evaluation of the Textbooks in Terms of STS

The textbook is the main source of study material for the student and the teacher. This is true for science textbooks, which emphasize hands on activities, especially at the school level. Various science education teachers and researchers have evaluated science textbooks for different aspects in science. The textbook should portray the science content in a more personal and social manner. The studies in this category deal with the analysis of the textbook in terms of their coverage of STS issues.

Fensham (1976) defined three dimensions for social content of science education: (a) social nature of scientific work, which included development of science as a result of inputs by a number of people over time and across the world; (b) social application of science, where scientific knowledge was seen as being applied in the personal and social life of all citizens over the world; (c) 'Social ideology' which considered the value aspect of science, i.e. if an application was known or seen to have desirable outcomes, it would be considered positive; with bad consequences, it would be considered negative. Fensham content-analyzed some textbooks of UK, USA and Australia to determine the relative extent to which the three dimensions of social content were presented. All the text materials omitted the social nature dimension. There was considerable variation regarding social application of science, from hardly a mention in the "Nullified Combined Series" (UK), to many examples given in "Environmental Science" (UK). However, most examples of applications were industrial ones. Whereas the text of earlier years presented only the useful applications, later material gave both positive and negative applications. Science textbooks differed in emphasis given to various dimensions of the science and society inter-relationships. Fensham also compared the questions present in the textbooks with the idea that the type of question asked emphasizes the portions worth learning.

Levin and Lindbeck (1979) analyzed treatment of eleven moral and social issues, namely Darwinian evolution, diseases, drugs, environment, human genetics, human reproduction, man—his place in nature, origin of life, the population explosion, biological effects of radiation and race-variation in man. The term "issue" referred to a matter disputed by society. Several biology teachers validated the criteria for defining categories. The reliability of the categorization was measured as the Pearson Product–Moment correlation between rating scores of two science educators. The correlation ranged between 0.71 and 1.00, indicating reliability for the method. The authors concluded that textbooks differed significantly in the presentation of controversial issues and that no single textbook gives adequate coverage of all the issues. Thus, teachers should not depend on any one book as a source for study of social issues but use their own initiative to enrich the curriculum.

Souque (1987) studied the STS in thirty-two science textbooks used in primary and secondary schools across Canada. The thirty-two available texts that contained messages about STS interactions were classified into four groups: (i) application of knowledge (ii) socio-economics (iii) politics and (iv) scientific responsibility. The results of the study revealed that the application of knowledge category showed few or no messages; a total of 230 messages appeared regarding category two, socio-economics; 130 in political category, and nearly all texts failed to place scientists in their historical or social content. Only in five out of a total of twenty-eight textbooks, at least half the time scientists were mentioned, used a scientist's achievements as an opportunity to present in a systematic way the methods of science. Nearly all texts failed to place scientists in their historical or social context. Rarely were students told in any of the texts surveyed, for example, that Rutherford, Bhor, Thompson, and Chadwick not only knew each other but even worked at various times within the same laboratories. The purpose of the study was to identify technology and societal items that could be related to and/or included in a general science core curriculum. In particular, it had hoped that knowledge of these items would help to reorient the Malaysian form IV and V general science curriculum towards an emphasis on the interactions between science, technology and society. The result of such an orientation would enable a great majority of the non-science students taking the course to become more aware and appreciate the relationships between science and technology and their impact upon the individual and society. A general conclusion was that this is a useful method for identifying STS content suitable for inclusion in a general science core curriculum. Science teachers holding a B.Sc. degree and teacher certificate are suitable participants for a study of this nature. It was suggested from the results of this study that: (i) The list of STS topics identified in the study should be considered for inclusion in the general science core curriculum, or as a base for STS curriculum. (ii) Plans should be made to implement in-service courses for teachers responsible for teaching the course. (iii) STS topics resulting from this study should be reevaluated and revised periodically. (iv) Terminal examination should be changed to accommodate the aims and objectives of the curriculum. (v) Further studies should be undertaken to determine the specific needs of rural *versus* urban needs in STS curriculum.

Rodger Bybee (1987) studied what kind of topics should be covered in an STS course in science education or selected for insertion in an existing science course. He explored this question with 262 science educators from forty-one countries by the spring of 1984. The top twelve issues that emerged from this survey, in order of perceived importance, were: (i) world hunger and food resources, (ii) population growth, (iii) air quality and atmosphere, (iv) water resources, (v) war technology, (vi) human health and disease, (vii) energy shortages, (viii) land use, (ix) hazardous substances, (x) nuclear reactions, (xi) extinction of plants and animals, and (xii) mineral resources.

Roy (1988) analyzed high school general science textbooks with respect to objectives, content accuracy, relevance, and needs of students and teachers. All persons connected with the preparation and use of textbooks (authors, illustrators, editors, teachers, students and school heads) were sampled. It was determined that the textbooks analyzed were written according to the requirements of the curriculum.

Ramanathan (1993) conducted a study on an assessment of the representation of science in the NCERT upper primary science textbooks. The objectives of the study were to identify the various aspects of scientific literacy which are being represented in science textbooks and to qualify the relative emphasis given to the various aspects. The findings reveal four aspects of scientific literacy that have been identified to assess the textbooks. These are (i) science as a body of knowledge, (ii) science as a way of investigation, (iii) science as a way of thinking, and (iv) the interaction of science, technology and society. Scientific literacy had been chosen as theme for assessment. NCERT science textbooks of the upper primary level of schooling (class VI-VIII) had been selected for assessment. The second set of books for comparison was the Headway Science Series. The methods of content analysis had been chosen for quantification of the aspects outlined. The units identified in the textbooks and used for analysis were paragraphs, illustrations, tables and questions. The first three aspects were given 60 per cent, 15 per cent and 1.57 per cent emphasis respectively. Less attention was given to the social content of science, as also to the positive and negative effects of sciences and career opportunities in science and technology.

Soong and Yager (1993) studied the inclusion of STS material in the most frequently used secondary science text books in the USA. The major purpose of this study was to examine eleven of the most frequently used secondary science textbooks in US schools to determine the inclusion of STS materials as defined by Piel (1981) in his framemark for defining STS as special topics. Such an analysis is viewed as critical because (i) STS has been defined as a major goal for science education, (ii) Most features use textbooks extensively—often following them page by page. The major question considered was how do the eleven books compare as to the treatment of STS topics? Other questions derived from this major question include the following: (i) how much space, in terms of percentage of the total narration included in each textbook, is devoted to the STS topics, (ii) what are the differences from textbook to textbook in STS treatment for a given discipline, (iii) how do the textbooks vary across the grade level with respect to attention given to STS? The findings were that energy was the only topic covered by all of the textbooks examined across grade levels and disciplines. However, this coverage was no more than two per cent of the total narration of each book. Environmental quality/natural resources and effects of technological development were presented in all but two of the textbooks. Across grade levels, human engineering and space research/material and defence secure the least coverage among the STS topics. This was less than one per cent.

The studies conducted on textbook evaluation in terms of STS showed that the text material omitted the social dimension. Textbooks differed significantly in the presentation of controversial issues. Teachers using them should use their initiative to enrich the curriculum. STS topics should be identified for inclusion in the general science core curriculum, these topics should be revised periodically. Less attention was found to be given to social context of science, as also to the positive and negative effects of science and career opportunities in science and technology. In terms of percentages less than two per cent content was devoted to interaction of science, technology and society. More studies are required for special needs of rural versus urban needs of STS curriculum.

Studies Related to Beliefs and Perceptions of Students, Teachers and Science Educators about STS

Education for scientific and technology literacy is grounded in the need for the citizen to understand and make decisions about the limits and possibilities of science and technology in resolving personal and social issues and achieving cultural aspirations. It has been argued that for the students to become active and informed decision-makers of science and technology related societal issues science education should promote an understanding of the nature of science and technology and their interaction within society. Researches have been carried out to know the impact of STS education and the students' view about the STS interactions. Since science teachers and science educators are the ones with the specific task of clarifying what future citizens need to know and decide upon, what the important personal needs and social problems are, and finally, how the work of science and science teaching relates to cultural aspirations, it is important to know how they perceive S-T-S. This section covers the beliefs and perceptions studies related to the students, teachers and science educators about STS.

Studies Related to Students' Beliefs and Perceptions About STS

Bybee, Harms, Ward and Yager (1980) in the paper 'Science, Society, and Science Education' discuss the students, attitudes based on the National Assessment of Educational Progress (NAEP) 1976-77. Questions were administered to probability samples of nine, thirteen, and seventeen-year-old students representing the national population of students at these age levels. 2,500 students responded to any one item.

This chapter explores the relationship between science, society and science education. Several important components of this relationship were considered. First science educators must understand the trends and issues associated with the public perceptions of science, second, people need to be aware of science related 'social' problems and should realize that their personal decisions help make up large social decisions that can either perpetuate or alleviate social problems, third, public awareness of scientific methods and values is essential in our society, finally, if social institutions, such as science, are to help ameliorate societal problems, there must be confidence in and support for, the institutions. The data concludes that science education is not preparing students for their role as citizens who must make informed and responsible decisions about science related social problems. NAEP data indicates that students are aware of what they can or should do about science related problems, but that their behaviour does not necessarily reflect this awareness. This can be done within science classrooms, for example, paper, electricity and water can be consciously conserved in science classrooms. There is need for education about the basic nature of science related problems such as energy, pollution, population, food supply, disease, resource conservation and human behaviour. Attention should be directed to exploitation of controversial issues. The NAEP data also suggests areas for further research on students' current information about societal issues and science. It would be useful to know whether students obtained information in the home, from the media, how they rank the relative importance to various science-related social problems and how they would justify their attitudes or positions.

Fleming (1986) studied adolescents' construction of science-related social issues and social cognition. Thirty-eight adolescents who had completed at least one year of biology and chemistry were selected for the study. Two STS issues, nuclear power plants and genetic engineering, were presented to the students and they were asked a series of questions. A large number of statements from students concerned scientific terms but students uniformly denied that science classes were sources of useful science information. When questioned how scientists *versus* non-scientists would stand on these two STS issues, subjects agreed that non-scientists would be against both developments. Adolescents did not perceive scientists as ordinary human beings. Science curricula must present science as a product of human endeavours with a strong emphasis on the human. The personal and emotional commitment to the creation of knowledge must be presented.

Fleming (1987) studied undergraduate science students' conceptions of technology. The purpose was to study views of undergraduate chemistry students on the nature of technology in the light of our theoretical understanding of the same and its interactions with society. The student views were examined using a variation of the views of science, technology and society instrument, (VOSTS CDN-2 Aikenhead, Fleming and Ryan, 1987) as well as an interview with selected students. The findings of the study revealed that undergraduates studying science do not know what technology is.

They equate technology with the production of artifacts, particularly those created by engineers. They do not perceive technology as having an organizational component (the administration of technology policy, the need of unions and consumers of technology) or a cultural components (issues of ideology such as progress, habits of thinking in technical issues, ethical codes).

Tamir (1987) studied the Israeli students' attitudes towards an understanding of the relationship between science and society and compared the Israeli data with corresponding data from other countries which participated in the second IEA Science Study. Data were collected from senior high school students in Israel within the frame work of SISS. A variety of additional data were available which were used to find out how various factors are related to the data on science and society. Some information about science and society was available for ninth grade and fifth grade to identify developmental trends. Lastly, a letter was sent to the twenty-five countries participating in SISS requesting them to provide the raw data on the pertinent items from the three instruments (attitudes, understanding, experiences) for international comparisons. The study revealed that attitudes towards science and society are positively but weakly correlated with achievement. Parent's education and the number of books at home are positively correlated with science career. There are no difference between boys and girls in regard to science as not harmful or as relevant to everyday life. Only students who did science studies in the upper secondary school have substantially less positive attitudes towards science and society. Students who have more positive attitudes towards science and society tend to choose science as a field of study in the university and eventually also as a career, especially in engineering and natural sciences. The situation is even worse in other countries with the exception of Finland and Papua New Guinea. In the Netherlands the ratings are the lowest in both attitudes and learning experiences. The high ratings of learning experiences related to science and society in Finland and New Guinea are not reflected in the students' attitudes.

Aikenhead, Fleming and Ryan (1987) studied the high school graduates' beliefs about STS materials and issues. The main purpose of the study was (i) to monitor student beliefs about STS topics by use of a large national sample of graduating high school students and (ii) to do this in a manner that improves upon similar search projects mentioned above and in a way that could lead to a new generation of standardized instruments. Sampling for this study was carried out by the Canadian IEP study. A stratified sample was selected. VOSTS instrument was used to analyze students' response. The STS topics included for consideration were those emphasizing attitude. A file of responses was organized for each of the 46 statements. The findings revealed that the ability of traditional instruments to capture students, belief is challenged by the results obtained from analyzing paragraph and a VOSTS "student position" is much greater than the precision of communication between a student's paragraph and his or her "agree" or "disagree" response. If one objectively scored the VOSTS responses one would sacrifice precision on this altar of quantitative methodology. Researchers who rely on students Likert type responses to, say, the Science Attitude Inventory (Mooree Sutmany 1970), the Science Process Inventory (Wuelch, 1966) or the Nature of Scientific Knowledge Scale (Rubba, 1976), run the risk of misunderstanding students' viewpoints. Revisions of VOSTS, or future developments of related instruments' could entail (i) a broader sampling of the domain of STS, (ii) an introduction of semi-structured interview component, or (iii) a multiple choice type of response format, using student generated positions as choices. Fruitful directions for further study include cross-cultural comparisons, comparisons among various groups of educators and pre-instruction investigators that would inform curricular developers and teachers who embrace STS goals.

Fleming (1987) studied high school graduates' beliefs about science, technology and society interactions. The questions that were studied were: What views do high-school graduates hold on the nature of the interaction of science, technology and society? What role do they see both science and technology playing in ensuring their quality of life? Should their society control the direction of both scientific research and technological development? If so, how would this be done? It reports on Canadian student responses to item 1 to 12 on the views on Science Technology Society (VOSTS). These items focus on the interaction of science, technology and society. The findings reveal that one major interaction between science and society was viewed by students in a rather simplistic fashion; Science (technoscience) should inform society in order to resolve socio-scientific issues, issues which students perceive as technical problems; but society should inform science in terms of science policy as it guides research programmes. The formulation of policy for a research programme was not perceived as socio-scientific issues.

Aikenhead (1987), studied high school graduates, beliefs about science, technology and society characteristics and limitations of scientific knowledge. The purpose of the study was to ascertain how literate high school graduates are when it comes to understanding the nature of scientific models, facts and methods, the tentativeness of scientific knowledge, and the social dimensions of scientific knowledge including both the academic and R&D milieu. It reports on Canadian Student responses to questions 13 to 19 of the VOSTS. The questions deal primarily with the characteristics and limitations of scientific knowledge. The study reveals that high school graduates harboured diverse and contradictory beliefs about scientific knowledge. Overall, the students argumentative paragraphs did not give the impression that students were generally confident in discussing the issues raised by the VOSTS questions. It was evident that students in the sample were not taught scientific modes, scientific method, the tentativeness of knowledge, and the R&D content of science. The data provide clear support for the renewal of science education towards teaching authentic science, and teaching with an emphasis on the science, technology and society connection.

Ryan (1987) studied high school graduates' beliefs about the characteristics of scientists. The purpose was to find how scientifically literate high school graduates are. More specifically, what do they believe about the way scientists view the effects of scientific discoveries? To what extent do they believe scientists are biased in their research work? What causes do they ascribe to the statistics showing that there are many more male scientists than female scientists? It is the fourth of a four-part set derived from a Canadian study. Each student was asked to respond to a statement concerning an STS topic by indicating whether he or she agreed with the statement, disagreed, or could not decide the choice. The findings reveal that, (i) the responses to view the relationship between scientists and their work suggest that on the whole students feel that scientists are being responsible in their actions, (ii) the responses to view on reporting of scientific finding to general public suggest that most respondents felt that scientists should be required to produce public reports, (iii) the students' views on the honesty and objectivity of scientists were varied, viz. scientists need not be honest, they should be honest, they are inherently honest etc., students' views and the unequal gender distribution of Canadian scientists suggested that the sensitivity and understanding exhibited by most male respondents to this question argue well for future initiatives to minimize biological socialization factors which are contributing to the low proportion of women in the science based professions.

Aikenhead (1988) did an analysis of four ways of assessing students' beliefs about STS topics. The study investigated the degree of ambiguity harbored by four different response modes used to monitor students' beliefs about science–technology–society topics. Likert types, written paragraphs, semi-structural interviews and empirically developed multiple choice were used. Grade-12 students in a Canadian urban setting responded in each of the four modes, to statements from views on science–technology–society. It was discovered that TV had far more influence on what students believed about science and its social, technological context than did numerous science courses. The challenge to science educators is to use the media effectively in combating naïve views about science. Regarding ambiguity in student assessment, the Likert type responses were the most inaccurate, offering only a guess at students' beliefs. Such guesswork calls into question the use of Likert type standardized tests that claim to assess students' views about science. Student paragraph responses contained significant ambiguities in about 50 per cent of the cases. The empirically developed multiple choices, however, reduced the ambiguity to the 20 per cent level. Predictably the semistructural interview was the least ambiguous of all four response modes, but it required the most time to administer. These findings encourage researchers to develop instruments grounded in the empirical data of student viewpoints rather than relying solely on instruments structured by the philosophical stances of science educators.

Aikenhead (1990) in the article "Assessment of students learning in the area of STS" describes a new methodology for developing test items which monitor students learning in the area of STS. It gives brief description of STS science (a curriculum perspective) followed by a brief summary of the evaluation problems out of which the new methodology arises. The new methodology for developing STS assessment items is then described in detail. The VOSTS project has developed a pool of approximately hundred items, which address a variety of STS topics. It is a reasonably accurate, efficient tool for assessing students' view points on STS topics. VOSTS offers accurate data and ease of administration for both classroom teachers and STS researchers.

Thirunarayanan (1990) conducted an exploratory multi-dimensional scaling of students' conceptions of selected science, technology and society issues. The main purposes of this study were to explore students' conception of thirty-one selected science, technology and society issues. Differences in such conceptions across different grade-levels and among males and females were also explored. A total of 138 males and females enrolled in grades five through eleven in a unified school district and in freshman-level courses in a large university constituted the sample for this study. The method of sorting was used to collect data on the thirty-one selected science, technology and society issues, from the subjects at all grade levels. The major findings of the study were, (i) the subjects were able to relate to issues which have personal implications, such as human health issues and transportation safety issues, by the time they were in the fifth grade, (ii) by the time they were in the ninth grade the subjects had developed meaningful conceptions of environmental issues which have local to global, as well as personal implications, (iii) issues related to automation of work caused concern among female subjects at the 9th grade and undergraduate levels, (iv) there was evidence of progression of students' conceptions of issues related to human health/disease and transportation safety at the fifth grade level and progressing to conceptions of broader environmental issues at the higher grade levels, and (v) a combination of qualitative and quantitative techniques of data analyses can be fruitfully employed to conduct research in the field of science, technology and society education.

Aikenhead (1990) studied as to what is the preferred testing format for assessing students' beliefs about science-technology-society topics. He compared the degree of ambiguity associated with four kinds of assessments used to monitor beliefs about STS topics. Twenty-seven twelfth grade students representing two Canadian high schools and a wide range of student achievement, responded to statements from the Views on Science–Technology–Society (VOST) in four ways: Likert–type "agree", "disagree" or "can't tell"; a written paragraph justifying personal reactions to VOST statements; a semi-structured interview; and the choice of STS positions empirically-derived from student paragraphs. From the most to least ambiguous, the four response modes were sequenced as follows: Likert-type statements, written narrative, multiple choice and interview. Although the interview generated the most unambiguous data, its liability of time prompted the author to recommend the use of the empirically-derived multiple-choice response mode which was found to be unambiguous about 80 per cent of the time. The Likert-type data provided little more than a guess about STS beliefs. However, the author was quick to explain that Likert statements are valid only for measuring attitude; VOST statements stress cognition. Aikenhead also sought to determine the source of the STS beliefs. Seventy-three per cent of the students cited the media as the source of their beliefs. Ten per cent cited science class, but no one mentioned science textbooks.

Ryan and Aikenhead (1992) studied students' pre-conceptions about the epistemology of science with the help of VOSTS. The study shows how VOSTS, a new research tool, documents students' reasoned beliefs about the following topics in the epistemology of science: scientific assumptions, values in science, conceptual invention, scientific method, consensus-making in science and characteristics of the knowledge produced in science. Because the choices for each VOSTS items were derived empirically from students' views (rather than from science educators' philosophical positions), a researcher can feel secure in the validity of student responses; that is, the meaning that students read into the VOSTS choices tends to be the same meaning that students would express if they were interviewed. The results were not encouraging; the image of science portrayed to students is both inaccurate and inappropriate. The results of the present study support the claim that if science instruction is going to convey accurate and appropriate images of science, it must be brought into line with contemporary epistemology.

Meichtry (1993) studied the impact of science curricula on student's views about the nature of science. Various interpretations of what aspect characterizes the nature of science were examined, revealing an agreement among scientist, science educators, and those involved in policy-making arenas that the nature of science is multi-faceted and an important component of scientific literacy. A summary of the research regarding the adequacies of students' conceptions about the nature of science revealed in consistent results. Although the majority of studies show that students' understandings are less than desirable, there is research that indicates that students' conceptions are acceptable. Research on the impact of the instructional materials and techniques on students' understandings was also reviewed. The effects of language in science instruction, the content emphasis of instructional materials, integrated science curricula and instructions in general were curricular variables found to have a negative impact on students' understandings about the nature of science. Empirical evidence about the success of innovative instructional materials and techniques designed to facilitate more adequate understandings of the nature of science is needed.

Bradford, Rubba and Harkness (1995) studied the views about science–technology–society interactions held by college students in general education physics and STS courses. This study examined whether and to what extent a general education STS and physics courses helped college students build more informed realistic views about the interactions among science, technology and society. Two samples took part in the study, one consisting of 138 college students enrolled in a general education STS course, STS 200–Critical Issues in Science, Technology and Society and the other comprising 122 college students enrolled in a general education physics course, PHYS 001–The Science of Physics, at a Large Eastern Land Grant University. Pre-test and post-test data were collected using 16 multiple-choice items selected from the 114 items in the Views on Science Technology Society (VOSTS) item pool. The findings for each of the 16 VOSTS items were treated descriptively. Additionally, a special scoring procedure was devised for the VOSTS items to allow the use of inferential statistics. The STS students moved towards more "realistic" views of STS interactions on a number of VOSTS items and toward more "native" views on other VOSTS items. The findings supported the value of general education STS courses. However, they also suggested that courses like STS 200 have room for improvement. The PHYS, 001 course had almost no impact on students' views of STS interactions. It should not be expected that college general education science courses, such as PHYS 001 will help students develop more adequate, realistic understanding of the STS interactions.

Jarvis and Rennie (1996) measured children's perceptions about technology using a writing/drawing activity and a picture quiz, administered to approximately three hundred grade three to six children in six primary schools in the English Midlands. The results indicated that although many children associate technology only with computers and modern appliances, overall there are a variety of ideas, with the more complex and coherent ideas among older children. Subsequently, interviews were carried out with 81 of the children to explore their reasons for choosing items which they said were something to do with technology. The children were able to respond in a thoughtful and consistent manner. Even when the children professed not to know what technology is, they often used an alternative but consistent explanation. They presented a tentative model describing the development of children's ideas about technology.

Rubba, Bradford and Harkness (1996) developed a new scoring procedure for the views on science-technology-society instrument. As originally developed and implemented, the VOSTS did not lend itself to test-retest comparisons and hypothesis testing using inferential statistical procedures. This paper describes the development of scoring through a re-analysis of VOSTS data from the initial application. If limitations associated with VOSTS item selection and categorization of VOSTS multiple-choices within the scoring scheme are attended to, the scoring procedure opens the possibility of applying inferential statistics to VOSTS data.

> In all the studies, students perceived that science education was not preparing them for their role as citizens to make informed and responsible decisions about science-related social problems. According to these studies, science should be presented as a product of human endeavor with a strong emphasis on the human. Not many students were aware about the word technology, they perceived technology as artefacts created by engineers without any organizational or cultural component. VOSTS has been described as an efficient tool for assessing student's view points about STS. The effects of language in science instruction, and the content emphasis of instructional materials were found to have a negative impact on students' understanding about the nature of science. Thus the combination of qualitative and quantitative techniques of data analysis as suggested to be fruitfully employed to conduct research in the field of science, technology and society education

Studies Related to Teacher's Beliefs and Perceptions About S-T-S

Bybee and Bonnstetter (1987) studied the perceptions of science teachers about implementing the S-T-S theme in science education.

A survey of science teachers was conducted concerning their perceptions of global problems, policy issues and their recommendations for curriculum and instruction relative to the STS theme. During the summer and fall of 1984, 317 teachers from across the world were surveyed. The sample group was 55 per cent male and 45 per cent female and had a mean age of 41. The sample included 27 per cent elementary teachers, 25 per cent middle/junior high school level and 41 per cent high school science teachers. The remaining 7 per cent were administrators or supervisors. Teachers were asked whether they consider incorporating science–technology–society theme into any aspect of their programme. A remarkable 89.2 per cent indicated they had considered the theme, which reassures that science teachers were indeed aware of the S-T-S theme in science education and they were conscious of the possibility of incorporating it into their programme. 90 per cent of them showed keenness in incorporating the S-T-S theme into their science programme if curriculum materials and instructional strategies were available. Out of 71.9 per cent responses received for the question as to how they would probably incorporate the S-T-S themes, 25.9 per cent wanted incorporation through single lessons or activities lasting one to three days. 23.3 per cent wanted short vignettes infused within lessons and lasting 3-5 minutes each and 22.7 per cent wanted a total K-12 S-T-S curriculum for the school district. A summary of those finding are: (i) science teachers rated air quality and atmosphere, water resources, population growth, energy shortages and world hunger and food resources as the top five S-T-S issues; (ii) science teachers knew least about war technology, mineral resources, world hunger and food resources, hazardous substances and land use; (iii) science teachers knew most about energy shortages; water resources, population growth, human health and diseases, and air quality and atmosphere; (iv) science teachers' primary sources of information about S-T-S were newspapers, television, professional journals, weekly magazines and books; (v) science teachers were generally pessimistic about the improvement of S-T-S issues by the year 2000; (vi) science teachers indicated that it is important to study S-T-S issues at all educational levels, with increasing emphasis in the higher grades; (vii) science teachers indicated their local schools are teaching "very little" or "some" about S-T-S; (viii) science teachers recommended the following as minimum amounts of instructional time on S-T-S themes, 6-10 per cent during elementary school, 11-15 per cent during middle/junior high school and 16-20 per cent during high school; (ix) science teachers indicated that courses in which S-T-S is a primary focus should be required of all students; and (x) science teachers recommend that the science and social studies aspects of S-T-S issues should be integrated and taught in one course. To summarize, a majority of science teachers had considered incorporating the S-T-S theme into their programmes. An even larger majority wanted to incorporate the S-T-S theme into programmes if materials and techniques were available.

Mitchener and Anderson (1989) studied the teachers' perceptive in developing and implementation on STS curriculum. It explores science teachers' perceptions of and consequent decisions about the development and implementation of a non-traditional science curriculum. It is a case study analyses and examines the perceptions of fourteen secondary teachers of a model Science/technology/society (STS) curriculum designed to promote scientific literacy. The purposes of the study were (i) to examine the beliefs and value systems that influenced teachers' perceptions of a new STS curriculum, and (ii) to repeat the factors the teachers perceived to be critical in the development and implementation of this STS model curriculum, factors that influenced the teachers' decisions to accept, alter, or reject the course. Whether teachers accepted, altered or rejected the course, they shared several clusters of values, beliefs and concerns about topics. While the depth of conviction varied among the three categories of teachers, nevertheless these matters were pertinent in all of them. There were five themes common to all three groups. These themes, which warrant consideration in the development and implementation of other STS curricula, are as follows: (a) concern over content, (b) discomfort with grouping, (c) uncertainties about evaluation, (d) frustrations about student population, and (e) confusion over the teachers' role.

Rubba (1989) conducted an investigation of the sematic meaning assigned to concepts affiliated with STS education and of STS instructional practices among a sample of exemplary teachers. Through this investigation the investigator sought answers to four questions relevant to the present status of pre-college STS education among a sample of science teachers recognized for their exemplary teaching. (i) What meaning do the exemplary science teachers assign to concepts associated with STS education? (ii) What percentage of science course time do the exemplary science teachers believe should be allotted to STS education at the secondary school level? (iii) What categories of STS issues do the exemplary science teachers integrate into science courses? and (iv) What instructional procedures do the exemplary science teachers use to integrate STS issues into science courses? The findings reveal that the exemplary science teachers were aware of the need to integrate STS into secondary courses and were making a good faith attempt given their limited understanding of the goal of integrating STS into secondary science and about how to accomplish that. For the most part, however, the teachers appeared to believe the concepts and topics of a particular science course. Still, even if the teachers held the social responsibility perceptive, the instructional procedures the teachers used and time actually spent on STS, limited their STS instruction to issue awareness. The promise of a citizenry, literate and participatory on science and technology related social issues, is attainable, but only if science teachers develop and learn to apply effectively pedagogical knowledge, skills and effective qualities appropriate to STS education.

Zoller *et al.*, (1990) studied the case of British Columbia to judge the goal attainment in STS education and its reality. The objectives of the study were (i) to assess the impact of ST 11 on the view points of senior high school students concerning the complex topic "science, technology and (Canadian) society", (ii) to assess whether and to what extent some or all of the prescribed subordinate goals of ST 11 have actually been attained in the first year of province wide implementation, (iii) to obtain a preliminary data base from the field for possible future use in decision making and educational policy concerning STS courses and curricula, and (iv) to evaluate some selective aspects involved in the implementation of ST 11 under the particular local constraints of British Columbia. A questionnaire, comprising four representative statements selected form the VOSTS inventory from CDN-mc 4, constituted the research instrument for assessing the high school students' STS-related beliefs/positions. It was administered to 12th graders (N-473). The science course history and gender of the respondents were also recorded. Data collected provided the bases for the construction of an "STS response profile". The findings reveal that (i) a substantial change of the viewpoints/positions of senior high school students concerning the complex topic "STS", in the desired direction correlates with the exposure of student to the ST 11 course, (ii) some of the pre-specified superordinate goals of ST 11 which were typical for STES courses worldwide have been attained, albeit some of the related sub-goals had not been achieved, (iii) a simple straight forward assessment of the key outcomes of the ST 11 course is feasible and the data base thus obtained is meaningful, and very useful as the bases for decision making and future educational policy making concerning S/T/E/S courses and curricula, (iv) STS respond profile is contextually dependent and with respect to some STS issues, gender dependent, and (v) the VOSTS inventory is an efficient, useful instrument for assessing students' viewpoints/ positions on STS related issues.

Zoller (1991) studied teachers' beliefs and views on selected S-T-S topics and probed into STS literacy versus indoctrination. Objectives of the study were (i) to assess the beliefs and views of different groups of teachers on selected STS topics and to establish the corresponding baseline SBVPs (STS Beliefs and Views Profiles) for intra- and inter-population comparisons, i.e., between different groups of teachers and their students, (ii) to determine whether, and to what extent, the STS teachers constitute a "special breed" using their SBVP as the criterion, (iii) to detect gender differences (if any) in teachers' SBVP, (iv) to get a "hint" concerning the level of teachers 'STS' literacy, and (v) to obtain a preliminary data base from the field which will enable a tentative, first approximation research based response to the question: Is STS education leading to STS indoctrination? A questionnaire comprised six representative statements selected from the VOSTS inventory form CDN-mc. 4 Aikenhead (1987) constituted the research instrument for assessing teachers and student STS related beliefs and views. The questionnaire was administered to 103 teachers (144 M, 39F) at randomly selected schools and 149 Grade 12 students who took the STS 11 course. The findings reveal that, (i) there appears to be no difference between the beliefs and views on STS and non-STS teachers, (ii) Gender differences concerning some STS topics within the teacher population did exist to some extent, (iii) some minor differences between the SBVP, of STS science teachers and those of both STS and non-STS non-science teachers do exist, (iv) the "STS-literacy" of teachers, particularly concerning the nature of science and technology, the difference between them and their interrelationships is questionable and (v) in STS course the norm appears to be "STS education" rather than "STS indoctrination".

Rubba and Harkness (1993) examined the pre-service and in-service secondary science teachers' beliefs about S-T-S interactions. The purpose of this investigation was to examine and compare pre-service and in-service teachers' beliefs about STS, particularly beliefs about the nature of science and technology and their interactions with society. The beliefs about the nature of science and technology and their interactions within society expressed by the samples of pre-service and in service science teachers examined in this investigation were equivalent for all practical purposes. TBASTS results showed that large percentage of the pre-service and in service teachers in two samples held misconceptions about the nature of science and technology and their interactions with society.

Ahmad (1995) conducted a study on teachers' awareness about education for STS and their perception of existing science content at secondary level in achieving identified ESTS objectives. The objectives of the study were (i) to identify and validate the objectives of Education for Science-Technology Society (ESTS) in Indian context, (ii) to study the awareness of Secondary School Teachers about ESTS, (iii) to study the views of science educators (curriculum developer, teacher educators and professional researchers) about ESTS and (iv) to study the perception of science teachers of secondary school regarding suitability of text books in achieving identified objectives of ESTS. The sample consisted of fifteen science educators and one hundred and fifty teachers of secondary/senior secondary schools of Delhi. The data were collected with the help of interview schedule, questionnaire and TBA-STS (Teachers' beliefs about STS). The findings of the study were that, (i) about 9 per cent of the content is devoted to STS material in the NCERT text book of science for class ninth, (ii) about 16 per cent of the content is devoted to STS material in the NCERT text book of science for class tenth, (iii) only a few things of STS-like nature of science, history of science and technology, social implication of science and technology, science as a social enterprise, inter-relationship of science and technology, relationship of science and society and science and technology as a product of culture are highlighted in the books. There were no clear interactions of science-technology-society highlighted in the text books, (iv) STS themes like—use of scientific method in personal and social life, philosophy of science, understanding and differentiating the hypothesis, theories and laws of science, nature of technology, moral ethical creative and humanistic dimension of science and technology, the societal control over science and technology, the difference between technology education and vocational education and the dynamic interactions between science–technology–society were untouched in the text books, (v) many covered themes like nature of science and history of science, the societal influence on science and technology, the relationship between science, technology and society was not discussed properly, (vi) social implication of science and technology was found to be a dominant theme throughout the book of class tenth, (vii) the education for technology was not touched in the books, (viii) experts opined the lack of following components of STS in the text books of science for classes nine and ten: (a) nature of science and its various components, (b) history and philosophy of science and technology, (c) nature of technology, (d) limitations and scope of science and technology, (e) application of principles of science and technology in solving current social issues related to science and technology, (f) social, political, ethical, moral and humanistic dimension of science and technology, (g) scientific method and its use in individual and social problems, and (h) multifaceted dynamic interactions of science-technology and society, (ix) 38 per cent of the teachers have wrong notion about STS, (x) 17.2 per cent of total teachers had an average understanding about STS, (xi) 45 per cent of the teachers had above average understanding about STS, (xii) not even a single teacher could score highest possible marks, and (xiii) the frequency distribution of 134 teachers on TBA-STS was slight positively skewed (.123) and lepto kuritic (.242).

Above mentioned studies revealed that teachers believed in incorporation of STS theme in their science education programme if curriculum materials and instructional strategies were available. Teachers were uncertain about the evaluation procedures for STS curriculum. They were aware of the need to integrate STS into secondary courses. They wanted more clarification about STS and suggested in service training based on STS to be given to them.

Studies Related to Beliefs and Perceptions of Science Educators About S-T-S

Bybee and Mau (1986) conducted an international survey of science educators on science and technology related global problems. This survey evaluated one aspect of the Science-Technology-Society theme, namely, the teaching of global problems related to science and technology. Two hundred sixty-two science educators representing 41 countries completed the survey. Response was 80 per cent, findings included a ranking of twelve global problems (the top six were; World Hunger and Food Resources, Population Growth, Air Quality and Atmosphere, Water Resources, War Technology, and Human Health and Disease). Science educators generally indicated the following, (i) the science and technology related global problems would be worse by the year 2000, (ii) they were slightly or moderately knowledgeable about the problems, (iii) print, audio-visual media and personal experiences were their primary sources of information, (iv) it was important to study global problems in schools, (v) emphasis on global problems should increase with age/grade level, (vi) an integrated approach should be used to teach about global problems; courses including global problems should be required of all students, (vii) most countries are in the early stages of developing programmes including global problems, (viii) there was a clear trend towards STS, (ix) there was public support for including global problems, and (x) the most significant limitations in the implementation of the S-T-S theme (in order of significance) are political, personal, social, psychological, economic, pedagogical and physical.

Bybee (1987) studied the views of science educators on teaching about STS. It presents responses of science educators in the United States to a survey of global problems related to science and technology. Survey responses clarify several issues within the current debate and, more importantly, provide a basis for answering some basic questions about the S-T-S theme and school science programmes. Questions addressed in this article include; Which S-T-S problems are important? How will these S-T-S problems change by the year 2000? How many of the science educators know about S-T-S problems? How important is it to study S-T-S problems in school? Who should take courses in which S-T-S problems are presented? How should S-T-S courses be presented? How much emphasis should be placed on S-T-S problems? And where is the United States in the development and implementation of the S-T-S theme in school programmes? Findings resulting from this survey include the following (i) Science educators ranked population growth, water resources, hunger and food resource, air quality and atmosphere, and war technology as the most important S-T-S issues, (ii) Science educators indicated that most social problems related to science and technology will be worse by the year 2000, (iii) Science educators were most knowledgeable about population growth energy shortages and water resources, (iv) Science educators were least knowledgeable about mineral resources technology, hazardous substances, extinction of plants and animals, nuclear reactors, (v) Science educators indicated that elementary and middle schools are teaching very little or nothing about S-T-S issues, (vi) Science educators indicated that it is fairly important to study S-T-S issues at the elementary school, (vii) Science educators indicated that it is important to study S-T-S issues in high school, college, and in adult education, (viii) Science educators indicated that the minimum percent of instructional time emphasizing S-T-S issues in elementary, middle and high school should be 10 per cent, 15 per cent and 20 per cent respectively, (ix) Science educators indicated courses in which S-T-S issues were present and should be required of all students, (x) Science educators indicated that science and social studies aspects of S-T-S issues should be incorporated into one course, (xi) Science educators indicated that implementation of the S-T-S themes was just beginning, and (xii) Science educators indicated that trends relative to teaching about S-T-S issues had progressed from 'not-prominent' in the late 1950s moderately prominent in the mid-1980s and would continue until it is very prominent in the last years of this century. The views of science educators in the United States supported the contention that the science-technology-society theme is established in contemporary science education. Because the S-T-S theme is a part of science education, it was clear that there were implications for curriculum and instruction, teacher education and developing public support for teaching about science-technology-society in school programmes.

Waks and Barchi (1992) studied the perception of selected leaders on STS in US school science and their implications for STS education. This study explores with STS leaders, their perceptions of the status of STS and their plans for leadership activities and whether these were sufficient to advance the movement beyond what it has achieved to date. Sixteen individuals were identified who had distinguished themselves as STS education leaders within the Science through Science, Technology and Society (S-STS) Project, Technology and Society Learning Materials Project, National Technological Literacy Conference (TLC) to disseminate the STS idea project, the STS Network or the National Association for STS representing leadership roles in mainstream educational agencies (teacher-leaders, state supervisors, university-based science educators and national project leaders) and geographical regions (East, Midwest, West, Pacific). It was assumed that this top leadership group could give a snapshot view of the thinking of STS leaders across the nation. These leaders were asked to reflect on (i) what had already been achieved in the STS movement in science education and what remained to be achieved, (ii) how their own thinking about the STS innovation had matured, (iii) what they thought leaders should be doing at this stage in the innovation process, and finally, (iv) what they personally planned to be doing in the midterm future to advance STS education in the school. Data was collected on a free response questionnaire. The leaders agreed that, though the STS movement is viewed as an indication of new directions in science education, it has not succeeded in spreading a single authoritative definition or vision to STS education. In addition they believe the movement has failed to affect a deep change in the perceptions of science teachers about the goals of science instruction. They further suggested that in order for conceptual understanding to occur, teachers must have an opportunity to rethink their own understanding about modern science and technology. The content, which they select, must provide illustrations of science, technology and the lives of people in society, which are clear to them. This, in turn, reinforces clear conceptual definition of these terms. Knowledge content must provide clear maps of the important interactions between science, technology and activities in social and political institutions, which bind the issue. Clearly, the persistence of the science and society theme in science education for at least half a century and the shared perception that STS in its current form has penetrated the awareness of teachers indicates that a shift in the direction of science education is taking place.

> Studies revealed that science educators also believed that it was important to study global problems in school. They believed that STS theme is a part of science education and STS should form a part of school programmes. The leaders agreed that though the STS movement is viewed as an indication of new directions in science education, it has not succeeded in spreading a single authoritative definition or vision to STS education.

Studies Related to the Effectiveness of STS

Layton (1986) studied the STS courses for their problems of implementation in school system and suggested that the STS topics should not be controversial. Teaching about STS issues should be restricted in schools to the upper levels of secondary institutions. It should be flexible and liable to be changed with time and advancement. More hospitable environments are to be found, in out-of-school and non-formal text in the programme of adult scientific literacy.

Rubba (1987) in the study "perspectives on S-T-S instruction" shared the experience at the Penn State Center for Education in Science, Technology and Society in working with school districts and individual science teachers, the study showed three ingredients are necessary if STS is to be integrated into the curriculum—first, school administrations, teachers and significant number of parents might come to understand the need for students to be able to make responsible decisions on the STS issues and to take action in accordance with those decisions; second, teachers and supervisors (e.g. principals) needed to receive in service training in the areas of (i) the nature of STS, including the characteristic interactions among teacher and society, (ii) recent scientific and technological developments which may have society implications, (iii) the variety of instructional strategies which can be used to teach STS apart from existing STS curriculum materials, and (iv) the selection and appropriate use of extant STS curriculum materials. Third ingredient was an opportunity to practice STS instruction in a supportive environment. The inferences, which followed, for STS instruction from the NSTA, recognized that exemplary STS programme and the recent EE (Environment Education) literature were clear. If we desire students to possess the capabilities to act in a responsible manner on science and technology related societal issues, STS instruction must rise beyond the issue, awareness level to include activities which build competence in issue investigation and action, and include experience in the application of these to STS issues. Only then will future citizens be prepared to contribute to the resolution of the endless variety of science and technology related societal issues which confront our society.

Yager *et al.* (1990) tested students in 300 Iowa classrooms, grades four through nine. One group experienced traditional science and another experienced science with an STS focus. The classes were compared on five domains of science education, connections and applications, attitudes, creativity, process skills, and science content. Students exposed to an STS experience were superior to students in traditional science courses on the following outcomes: ability to apply information to other situations; attitude towards science, science instruction and science teachers; creative behaviour, and ability to perform basic science process skills. Students in the STS course also acquire an equivalent amount of science content knowledge.

Zoller (1991) conducted a quasi-quantitative study focused on science teacher education about teaching/learning styles, performance and students' teaching evaluation in S/T/E/S (Science-Technology-Environment-Society). An S/T/E/S-oriented ISMMC-IEE (Integrated Subject-Matter/Methods Course and Integrated Electric Examination) combination model of instruction was implemented in two specially designed undergraduate courses and one graduate course within college science teacher training programmes. These three courses served as case studies for class-based, quasi quantitative pilot investigation aimed at gaining a deeper insight into some of the issues involved in the implementation in college of non-traditional, open-ended, problem-solving-oriented teaching strategies which are in dissonance with the cognitive or effective styles functional paradigms of most students. This study into the dissonance issue revealed that prospective teachers were capable of handling the new instructional model and do gain in their higher-level cognitive learning. However, undergraduates perceive these courses to be either difficult or not in accordance with their needs, and their appreciation of the instructional technique and style employed is different from that of graduate students accordingly. The study suggested that although the ISMMC-IEE model was useful in S/T/E/S-oriented courses in science teacher training programmes, special attention to the implementation stage was required to close the gap between students' and S/T/E/S educators' functional paradigms.

Iskandar (1991) conducted an evaluation of the S-T-S approach to science teaching. The purpose of the study was to evaluate the effectiveness of science/technology/society (STS) approach as a reform in science education. The study involved twelve 'lead' teachers from the IOWA Chautauqua programme who agreed to participate and one hundred and nine students from grade six through grade nine. Each teacher selected two classes for the purpose of this study.

One class served as the experimental group and the other one served as the contrast group. The assignment of students in both experimental and contrast groups was accomplished randomly by the counsellors and administration of each school district. Pre-test and post-test in five science domains were administered to students in both STS and textbook oriented (contrast) sections. Analysis of variance, t-test, and analysis of covariance procedures were used to compare the results between the STS and the textbook-oriented approaches. STS students were as capable as textbook-oriented students in mastering many basic concepts; were able to apply basic science concepts in new situations significantly better; developed more positive attitude towards science, science teachers and science related careers significantly better; perceived the power of asking questions in science classes significantly better; perceived the power of problem solving in science classes significantly better; perceived the power of teacher strategies in science classes significantly better than textbook-oriented students. Female STS students developed more positive attitude towards science teachers and science related careers, significantly better than female textbook-oriented students. All analyses were made at 0.05 level of confidence. It can be concluded that the STS approach is as effective as the textbook-oriented approach for helping students apply science concepts; for developing student positive attitude towards science, science teachers, and science related careers, for inculcating student skills creativity; for helping students understand and use science process skills. In addition, the STS approach enhances positive attitude towards science for female students.

Edward (1991) conducted an analysis of students' scores in five domains of science education for concordance across domains and growth within domains related to ability of teachers to posing application items during an S/T/S interventions. The study considered the relationship between student achievement in the domains of science education and their teacher's ability to pose new and challenging applications. Additionally, student scores were compared across domains to substantiate the need for generally expanding student assessment in the classroom. The study involved 511 students of grades 6-12 taught by 18 teachers in 27 classes. The classes were divided into three study groups based on the application-posing ability demonstrated by the teachers as identified through analysis of their test results. All teachers conducted a four to ten weeks STS unit and collected pre- and post-test scores in the following domains: concept, process, application, creativity and attitude. Data was analyzed using analysis of covariance to identify effects application-posing ability of teachers in each of the domains and the interactions with student gender, teacher gender and even grade level. Further, post-test scores of students in three domains were correlated using Pearson correlation to demonstrate consistency in student scores across the domains. Results showed generally higher achievement by students on teacher made tests in classes taught by teachers demonstrating the highest level of application-posing ability. In the concept, process and application domains, the differences were significant at $p<0.01$. Generally there were no significant interactions between groups based on either student gender or teacher gender. Grade level interaction suggested that teacher abilities in application-posing are more significant for students in lower grades. While all domains demonstrated this trend, the only domain in which this variance was statistically significant was the concept domain. Analysis of students' post-test scores in the domains of concept, process and application showed much variance across domains. The Pearson correlation matrix produced no correlation above 0.1.

Rubba, Mc-Guyer and Wahlund (1991) studied the effects of infusing STS vignettes into the genetics unit of biology on learner outcomes in STS and genetics. The purpose of the study was to test the effects of infusing STS vignettes into the genetics units of high school biology instruct with periodic class discussion of the vignettes, as compared to not doing so on student outcomes at the lowest two levels of STS instruction. The three relevant student outcomes selected were (a) awareness of current STS issues, (b) the importance assigned to current STS issues, and (c) genetics achievement. Three directional null hypotheses were posed. The non-equivalent control group quasi-experimental design was employed. Four sections of grade ninth or tenth general biology were randomly assigned to the experimental and control treatments. STS vignettes were not used in the control classes. In all the students in the four classes were taught genetics for six weeks and were pre-tested and post-tested. Findings of the study revealed infusion of STS vignettes with periodic class discussion into the genetic unit of biology instruction as opposed to not doing so: (i) It did not result in a difference in student genetics achievement. (ii) It did not result in a difference in student awareness of current STS issues. (iii) It did not result in a difference in the perceived importance student assigned to current STS issues.

Yoshisuke (1993) studied the effects of STS instructions in Japan compared to results reported in the US. An experiment similar to those undertaken in Iowa was conducted to permit a comparison of results. Two Japanese teachers taught multiple class sections where both an STS and a traditional mode were utilized. Every two weeks, video recordings were made of the STS and traditional sections. A study of the video recordings provided a record of the two teaching behaviours used and the quality of STS teaching. The analysis of student's achievement in multiple domains permitted a comparison of Japanese results with those repeated in Iowa. Specific results with students learning in STS and traditional classes in Japan permit the following statements: (i) STS students do as well as students in traditional classes with respect to concept mastery, (ii) STS students in the grade ten samples developed significantly more positive attitudes about science study, science teachers and science careers than students in traditional sections, and (iii) STS students showed significantly more growth with respect to creating skills in both the grades eight and ten.

Yager and Tamir (1993) studied the STS approach in terms of reasons, intentions, accomplishments and outcomes. STS effectiveness as an approach to science in grade four through seven was studied. Ten experienced STS teachers used both the STS approach and the textbook approach in two different classes. 63 teachers new to STS collected pre- and post-information. Results of successes with STS were compared between experienced STS teachers in each of the two modes, namely STS and textbook approaches, and between the experienced and new STS teachers. Pre and post-test designs were used for assessing students in each of the five domains of science learning, namely concept, application, process, creativity and attitude. The relative effectiveness of the STS approach was analyzed with the ANCOVA and t-statistics. The results of analysis show, (i) students who experience the STS approach are more successful in the five domains studied than those who experience the textbook approach, (ii) teachers new to STS succeed with the STS approach as well as the more experienced teachers, (iii) the STS approach, whether used by the experienced or the inexperienced STS teachers, is more effective in improving the achievement levels for both high and low achieving students than the conventional approach and (iv) the STS approach resolves problems associated with lower achievement and more negative attitude commonly reported for female students.

Yager (1993) in the article "Make a difference with STS-should we toss out the textbooks", described the national Chautauqua Inservice Programme in Iowa from 1983-86. The Iowa Programme focused on STS materials and strategies. The Chautanqua Model was seen as one that could affect teacher attitude and behaviour. Assessments of the model reveal some exciting information about conceiving success with STS programmes. Twelve lead teachers agreed to teach one section of the student with STS strategies and another from the regular textbook. Pre- and post-tests were administered to both sections at the beginning and the end of a semester. The results of the tests indicated that STS did not result in greater concept mastery after one semester (some argues that STS strategy results in a more long term learning because the learning comes from direct experience). However, the growth in the other four domains was significantly better (0.01 level of confidence) for the STS sections. Students who practised and used process skills were significantly more skilled than students who studied science in traditional ways. The attitudes for students in STS sections were significantly more positive, i.e. very different from the expected more negative attitude that occurs in typical science classrooms. Creative skills were enhanced enormously and significantly in STS classrooms over textbook sections. Students asked more questions and more unique questions, they could suggest more possible causes for given events and in more unique areas, they could predict more consequences and ones that were more unique than students enrolled in textbook sections. Finally, students in STS classrooms were far more successful (again, statistically significant at the 0.01 level) in applying concepts to new situations. The success with the Iowa Chautauqua Model and the STS approaches that characterize its use in Iowa was striking. The results provided strong evidence that the model has a power worthy of trial elsewhere. Further, the results of STS effort resulted in students with more features that coincided with started goals for school science. STS provided more students with the characteristics of the scientifically literate person.

Lu-yu-Ling (1994) studied the effectiveness of STS approach to science teaching in the elementary school. The objectives of the study were to, (i) describe the Iowa STS model along with the goals, (ii) characterize and give concrete examples of using the STS approach in the class room, (iii) report results of four studies that have assessed the outcomes of learning with STS instructor, and

(iv) identify some problems concerning the future use of STS in schools. The first purpose was reached by reviewing the definition of STS outlined by NSTA and practised in Iowa. The second purpose was accomplished by the use of teachers' descriptions of their work with their students. The third purpose was achieved by re-analyzing and summarizing results of the four studies that were carried out at the University of Iowa. The final purpose was achieved by discussing the implications as the first three purposes were met. The study described five domains of Iowa STS Model. Use of STS approach in classrooms brings about more effective teaching and learning. Using STS approach, there is a major change between the pre-test and post-test stages in the five domains identified. Some of the major problems identified concerning the future use of STS in schools were evaluation instruments and teachers' continuous experience with STS approach.

Parashar (1994) studied the STS approach in teaching of science and its implication for science education in India. The purpose of the study was (i) to develop an insight into STS approach of teaching, (ii) to identify and study specific conceptual, curricular frameworks and projects used in STS approach, (iii) to draw generalizations from various research studies, based on STS approach, and (iv) to suggest the feasibility of STS approach in Indian context. The investigator had adopted document analysis method in her study. This method involved study and examination of the documents. These documents included dissertation abstracts, research reviews, research journals, works and seminar papers. This was done to collect literature on STS. The study provided a description of STS approach along with conceptual, curricular frameworks and projects in STS approach. It listed research studies and articles on STS. The findings of the study were that STS is being successfully used in some countries. The difference between the previous science curricular improvement and STS is that STS stresses more on relevance. The conceptual frameworks on STS clarify the objectives of teaching through this approach. The curricular frameworks suggest how to prepare curriculum keeping the approach in mind. Most of the projects on STS were from USA. There were few from the rest of the world. There was no research study reported from India. In general, the research studies and articles suggested that there is scope for improvement in science education using STS approach. At present if STS is to be introduced in the Indian situation, it should be done by infusion in the present curriculum. A new programme based on STS can be prepared to be first tried out for pre-service training and later on in classrooms and for in-service teachers.

Pedretti (1994) in the study 'Action research in STS education: the road less travelled' addressed issues of curriculum development and teacher education central to STS adoption and implementation. In an effort to contribute to the action research literature, particularly in the context of STS education, an action research group in science, technology and society was established at OISE in September 1992 until May 1993. The group comprised the author and six science teachers from the Toronto area: one from the elementary panel and five from the secondary panel. These teachers shared common interest in STS education. The study set out to (i) provide systematic and continuous opportunity (through action research), for science teachers to reflect upon and examine their own theories and practices in the context of STS education, (ii) provide opportunities for science teachers to engage in STS curriculum development, (iii) learn about the way one does action research in an STS context (in other words, how does the idea work in practice?), and (iv) describe the effectiveness of a particular kind of practice-action research in STS education and curriculum development.

Pedretti and Hodson (1995) in the article "From rhetoric to action: implementing STS education through action research" outlines the arguments for an action research approach to the implementation of STS education, and describes some of the experiences and outcomes of an action research group of six teachers (one elementary, five secondary) based at The Ontario Institute for Studies in Education, Toronto. The teachers' initial views and their changing perceptions throughout the duration of the project (the 1992-93 academic year), are presented in terms of four categories: the nature of STS education; the development of curriculum materials; personal/professional development; collaborative work. The emergence of an additional complex theme involving reinforcement, extending horizons and building professional confidence is considered the most significant outcome of the project.

Price and Cross (1995) in the article 'Conceptions of science and technology clarified: improving the teaching of science', elaborates that development in the teaching of science in recent years, whether under the umbrella of STS (Science, Technology, Society) or the narrower introduction of technology in some forms, have failed to clarify precisely what is performed under these names and how they relate. The paper proposed a distinction which will make clear the purposes and precise nature of the theorizing process. Description-explanation is contrasted with know-how, the recipes involved in making and doing. The distinction is seen not only as a way of avoiding terminological argument, e.g. about whether technology is applied science, but more importantly, as helping make explicit the thought process involved. The distinction is illustrated by examining the work of Josef Von Fraunhofer on making achromatic lenses and of Boyle and others in spectroscopic analysis.

Kortland (1996) studied an STS case study about students' decision making on the waste issues. The study is about students' decision making on the waste issue in the physical science curriculum. A teaching unit limiting waste issue to household packaging waste was developed and trailed in grade 8 middle ability classes with students aged 13-14. The students' existing and developing decision making ability was assessed against the background of a normative model of the decision making process, with evaluating alternatives on the relevant criteria at its core. The research connected to the classroom trials shows that before teaching the unit students were able to evaluate alternatives on one (or more) non conflicting criteria. After teaching the unit the students' argumentations about a decision making situation have improved as far as validity and clarity of the criteria used for evaluating alternative is concerned, but has remained stable with respect to the limited range of criteria used. After an exploration of the reasons for these disappointing learning effects, the article ends with some reflections on the way in which a new version of the unit might be more effective in developing the students' existing decision making ability.

Pedretti (1997) in his paper 'Septic tank crisis: a case study of STS education in an elementary school' describes that STS is a desirable and commendable goal. However, many dilemmas arise when educators begin to seriously address issues of power, knowledge, decision-making and responsible action in the science curriculum. In particular, this paper is a case study of an elementary teacher named Julie involved in STS education with her grades three-four classes. The real-life event or issue that triggers the STS curriculum unit is the school's septic tank overload. The author describes that the guiding principles describe an issue-based or event-centred learning approach to STS education that identifies social responsibility as a primary goal, and inherently includes critical thinking, decision-making, action and moral reasoning. The elementary teacher, Julie maintained a science programme that deviated from what she describes as the usual 'prescribed, boring science classes'. She explored alternative strategies and projects with her students, integrated subjects, politicized students at a young age, and initiated appropriate action through an issue-based approach to STS education. Yet, at the same time, there was a cautionary note in her words. As she reflected upon her teaching and planning, she questioned the extent to which STS education can truly promote the goal of scientific and technological literacy, and the extent to which she could create meaningful and relevant experiences for students. Vigilance, integrity and foresight need to be exercised as students become engaged in decision-making and possible responsible appropriate action. The call for science, technology and society education is clearly defensible and highly applauded at both the elementary and secondary levels. STS education sought to enrich the experiences of students, sponsored creativity and imagination about alternatives, and provided power and freedom to examine and question social issues related to science and technology. At the same time educators must be realistic and consider the pedagogical and curriculum implications and tensions that inevitably arise for those engaged in STS education. The rhetoric is much more convincing and less problematic than the reality of STS education. In describing this case the author examined the potential power and problems of STS education.

Solbes and Vilches (1997) in their study describe the STS interactions in the teaching of physics and chemistry. The absence of science-technology-society (STS) interactions and its consequences in the basic teaching of science in Spain are analysed in the first part of this work. This article proposes the introduction of STS interactions in physics and chemistry classes in conjunction with the teaching/learning model of science as a research. When such interactions are not introduced, it was observed that the students have a vision of science that is removed from the world in which they live and are unfamiliar with the mutual relationships between science, technology and the natural, social environments in which they are immersed. Not only do a large number of textbooks fail to cover STS interactions, but the majority of teachers do not consider interactive STS aspects necessary, nor do they contemplate these aspects in instruction. All of this contributes to the lack of students' interest in physics and chemistry and their rejection of them as subjects. In the second part of the work, students of 16-18 years of age in the last 3 years of secondary education were surveyed and the result obtained was analyzed. These results confirmed that dealing with STS interactions in the classroom established science as something more complete and integrated in the students' environment. Students subsequently developed an improved comprehension and a more real image of these sciences, which allowed them to understand better the role of scientists and how they work. All of this generated positive attitudes towards the study of physics and chemistry and increased the students' interest in their study. Thus the results of this research make it clear that it is possible to transform the learning of physics and chemistry with the inclusion of STS activities, so that the students can build scientific knowledge. Likewise, students integrate essential aspects that affect the scientific activity and contribute to deepening and consolidating their own knowledge.

Thus STS has been effective for enhancing students' achievements, understanding and attitude towards science equation. It showed that if we desire students to possess the capabilities to act in a responsible manner on science and technology related societal issues, STS instruction in terms of activities should be strengthened. Studies showed that students exposed to an STS experience were superior to students in traditional science courses on a number of outcomes and STS classrooms were found to be more successful in applying concepts to new situations.

Conclusion

A close review of the related studies as presented in the foregoing pages leads to the following conclusions:

1. A sizeable number of empirical studies have been carried out to show the need for the science education curriculum to be inclined towards the STS curriculum. Most of the studies conducted abroad have shown discrepancy between the curriculum in schools and the changing, technologically oriented society. In the curriculum there is more emphasis on the rote memorization of information. A need was felt for greater attention to student's non-social and social construction of knowledge and organization of courses in a

social context rather than special disciplines. The researches emphasized on four goals for inclusion of science in every school programme, namely personal needs, societal issues, career education/awareness and academic preparation, Hurd (1981) [Hurd (1985), Penick and Yager (1986) and Baez (1986)].

2. The findings of the many researches emphasized on the need for a balance of science and technology with an emphasis on inter relationships between and among them, values and society at large, development of courses around persistent social problems that have underlying health, environmental and technological aspects. Need was felt not only to create new environmental education courses but also the development of instructional methods and materials which can be integrated into existing courses and other types of science and technology activities both in and out of classroom. A coordinated reform is required in the entire school curriculum to address without loss an urgent attention to science concepts and skills development [Bybee (1987), Khan (1987), Hart and Robottom (1990), Okebukola and Jegede (1990), McFadden (1991), Bingle and Gaskell (1994), Hansen and Oslon (1996), Pedretti (1996) and Aikenhead (1997)].
3. A number of textbooks were analyzed in terms of the social application, value aspect of science and social nature of scientific work. Findings of studies revealed that though materials omitted the social nature dimension of science, there was considerable variation regarding social application of science. In some books the text presented useful application whereas in some, both positive and negative applications were present. The textbooks were found to be differing significantly in the presentation of controversial issues. STS interactions were present negligibly and then also the application of knowledge was represented negligibly. [Fensham (1976) and Levin and Lindbeck (1979)].
4. Some studies showed that there was very little coverage of the moral and legal dimensions of scientific and technological activities. The textbooks mentioned scientists and their

achievements but failed to place scientists in their historical and social context [Souque (1987) and Bybee (1987)]. There was one study which analyzed science textbooks with respect to objectives, content accuracy, relevance and needs of students and teachers and found them to be in accordance with the requirements of the curriculum [Roy (1988)].

5. Studies were also conducted on the inclusion of STS material in the existing textbooks which showed that though STS is defined as a major goal for science education, less than two per cent of the total education was devoted to STS [Ramanathan (1993) and Soong and Yager (1993)].
6. Studies related to the students' beliefs and perceptions about STS revealed that though the students were aware of what they can or should do about science related problems, their behaviour did not reflect it. The studies conclude that science education is not preparing students for their role as citizens who must make informed and responsible decisions about science related social problems. It must present science as a product of human endeavour with a strong emphasis on the human. Some students did not know what technology was [Bybee, Harms, Ward and Yager (1980), Fleming (1986) and Fleming (1987)].
7. Studies conducted on the understanding of students about science, technology and society revealed that the attitude towards science and society are positive but weakly correlated with achievement. Students who studied science at the upper secondary school have less positive attitude towards science and society than those who pursue science to the higher levels. The students harboured diverse and contradictory beliefs about scientific knowledge and thus associated technology with computers and modern applications. Thus recommending a renewal of science education towards teaching authentic science and teaching with an emphasis on the science-technology-society connection [Tamir (1987), Aikenhead (1987), Ryan (1987) and Thirunarayanan (1990)].
8. A few studies discussed the development of test items to monitor students' learning in the area of STS. VOSTS (views

on science-technology-society) instrument was discovered to be the most appropriate to assess students' beliefs about science-technology-society topics [Aikenhead, Fleming and Ryan (1987), Aikenhead (1988), Aikenhead (1990), Ryan and Aikenhead (1992) and Rubba, Bradford and Harkness (1996)].

9. Certain studies about impact of science curricula on students' views about nature of science showed that effect of language in science instruction, the content emphasis of instruction materials, integrated science curricula and instructions in general were curricular variable and were found to have a negative impact on students' understanding about the nature of science [Meichtry (1993), Bradford, Rubba and Harkness (1995) and Jarvis and Rennie (1996)].
10. The studies conducted to study teachers' beliefs and perceptions about STS revealed that teachers know very little about topics related to war technology, mineral resources, world hunger and food resources, hazardous substances. They rated air quality and atmosphere, water resources, population growth, energy shortages, and world hunger and food resources as the top five STS issues. Teachers indicated the importance to study S-T-S issues at all educational levels with increasing emphasis in the higher grades. They were in favour of allocating instructional time to STS themes. 90 per cent of them showed keenness of incorporating STS theme into science programme. Few of them were apprehensive about their role and uncertainties about evaluation [Bybee and Bonnstellar (1987), Mitchener and Anderson (1989), Rubba (1989), Zoller *et al.* (1990), Zoller (1991), Rubba and Harkness (1993) and Ahmed (1995)].
11. Studies related to the beliefs and perceptions of science educators about STS were similar to that of teachers' beliefs. They were slightly more knowledgeable about problems, they were supportive of including global problems in the science curricula with age/grade level and were in a favour for implementation of the STS curricula. They emphasized the need to study S-T-S issues at the elementary school. Because S-T-S theme is a part of science education, it is clear that

teaching about STS in school programmes will have public support [Bybee and Mau (1986), Bybee (1987) and Waks and Barchi (1992)].

12. Many studies were conducted on effectiveness of STS. A few studies were conducted to study the STS courses in schools and the problems of implementation. Their findings showed that STS topics should not be controversial. The STS instruction must rise beyond the issue of awareness level to include activities which build competence in issue investigation and action, and include experience in the application of these to STS issues [Layton (1986), Rubba (1987), Zoller (1991) and Rubba, McGuyer and Wahlund (1991)].
13. Some studies were conducted to test the effectiveness in the real classroom situation. The students were compared on the domains of science education. Students exposed to STS experience were found to be superior to students in traditional science courses on various aspects in these domains. They were superior in the ability to apply information in other situations, attitude towards science, science instruction and science teachers, creative behaviour ability to perform on basic science and process skills. STS approach improved the achievement levels for both high and low achieving students. The Iowa Chautauqua Model of STS discussed in many studies has been successful and is suggested to be tried elsewhere. But findings of one study are contradictory which showed that there was no difference in students' achievement, students' awareness of current STS issues and the perceived importance student assign to current STS issues even when STS topics were infused into one unit of the textbook [Yager *et al.* (1990), Iskandar (1991), Edward (1991), Yoshisuke (1993), Yager and Jamir (1993), Yager (1993), Prashar (1994) and Lu-yu-Ling (1994)].
14. STS education is clearly applauded at both the elementary and secondary levels as indicated by various studies but at the same time they caution to consider the pedagogical and curriculum implications and tensions that inevitably arise for those engaged in STS education [Pedretti (1994), Pedretti

and Hodson (1995), Price and Cross (1995), Kortland (1996), Pedretti (1997) and Solbes and Vilches (1997)].

Past Studies and Present Research

An overview of the studies carried out in the past in this area reveal that there is a need to have curriculum with a balance of science and technology with an emphasis on inter relationships between and among them, values and society at large, development of courses around persistent social problems that have underlying health, environmental and technological aspects. These studies also suggest that to make STS feasible and more effective in classroom situation the following will be required. Firstly, textbooks should be properly designed to deal, with STS topics. All the aspects of STS education should be covered in the textbooks. Secondly, students' conceptions about STS should be very clear. Students should hold positive attitude to deal with STS issues. Thirdly, teachers should also be competitive and hold clear conceptions about STS. Fourthly, while proceeding towards STS education, various problems are encountered. To overcome these it would be advisable to develop instruction materials and methods which can be integrated into existing courses instead of changing them totally. These should be flexible, i.e. liable to be changed with the needs of the particular time.

More emphasis should be given to local environmental issues. Content suitable for STS curriculum should be identified according to the needs and surroundings of the students, i.e. beginning should be from some familiar, local issues as the emphasis area, where meaningful active participation of student is possible, which will be related to the scientific concepts. There is a need to incorporate such topics and matters explicitly and implicitly within STS learning materials and instructional strategies, which will help in shaping STS education for maximum utility. The present research work has tried to make an attempt towards this direction. Even the curriculum models proposed for construction of curriculum of STS education need to be tested in real Indian classroom situation. It is readily apparent that there are lacunas in the present curriculum material design, which can be removed by testing the efficacy of such proposed models by various researchers. As researchers will construct STS curriculum materials using these models, gaps in a knowledge base in these areas will begin to be filled. A programme of research that attempts to address some of these issues, especially if linked with major STS curriculum development projects, could advance our understanding of the issues in an efficient manner.

The analysis of the review of these studies showed that all the studies had been conducted abroad except three M.Phil studies, which have been conducted in India. These three studies which have been conducted have the research base which is fragmentary, this is due to the fact that there are less longitudinal studies in this area, which have been conducted in India. These studies are limited in their scope and do not lead at present to generalizations that are of wider acceptance. So far as India is concerned, education of science, technology and society is still almost an ignored area, although nobody can deny its credibility and relevance in the Indian context.

There is need to formulate and validate STS themes in science education in the Indian context at various stages of school education. The textbooks need to be analyzed in terms of these validated STS themes. Textbooks have been analyzed in India but in terms of other variables and much of the work is limited to NCERT textbooks. No such attempt has been made to analyze the textbooks of Haryana Education Board, especially in terms of the STS themes. Need was also felt to document the efficacy of these STS themes in science education by making incorporations in the already existing content of the science textbooks. Research was also needed to see the feasibility of the incorporations made in the real classroom situation and to judge the effectiveness of these incorporations in terms of the nature of change in children's achievement and understanding of science, technology and society. So far as the knowledge of the investigator is concerned no such effort has been made in India on the above mentioned aspects. The present research work tried to validate the STS themes in science education in the Indian context at the upper primary stage. The work further analyzes the textbooks of upper primary stage in terms of theses validate themes, makes necessary incorporations in the content of the textbooks and tries the feasibility of the incorporations made in the real classroom situation by observing the change in the achievement and the understanding of science, technology and society as a result of incorporation of the STS input.

4

Design of the Study and Development of Tools

Introduction

After the formulation of the problem and having an insight into the field of literature related to the problem under study, the immediate task of the researcher is to search out the most appropriate plan and strategy to obtain answers to research questions. The plan or strategy constitutes the research design which guides the researcher in the process of collecting, analyzing and interpreting data. Adequately planned and executed design help greatly in permitting us to rely on both our observations and our inferences.

The introduction of STS in schools is needed for an informed public to contribute intelligently and in a responsible manner to decision making in problems related to impact of science and technology on society and their inter-relationships. Moreover science being taught to the rural children is not relevant to the needs of the majority of the students but only aims at preparing them for future study of science at the university level. This is in spite of the fact that only a very small percentage of students reaches the university and, of these, only a fraction specializes in the scientific field. Most of the children in rural area are only able to qualify at elementary stage. Thus, it is essential to make science at elementary stage more relevant to their needs.

The primary purpose of the present study was to draw a framework of STS themes in science education and to analyze the science textbooks of upper primary stage (VI to VIII) of Haryana Education Board in terms of this framework and subsequently judge the effectiveness of STS input in terms of achievement and understanding of science, technology and society. Keeping in view the nature and purpose of the present study, the sample, the measuring instruments for data collection and statistical techniques to analyze the data were decided. The present chapter deals with the descriptive plan and procedure adopted in carrying out this investigation, i.e. the procedure followed in selection of the sample, description of the variables, procedure of development and selection of tools, collection of data, scoring and tabulation of data and statistical techniques used to analyze the data.

The present study dealt with the renovation of science curriculum. It is worthwhile to overview the definitions of the term curriculum. Etymologically 'curriculum' is derived from a Latin word 'carrere' which means to run. Therefore, technically curriculum is a course which runs to reach a goal. But this meaning of curriculum has changed now, though this old-time meaning has not entirely vanished. Subject curriculum follows a fairly common formula which includes (i) expert judgement to determine the subjects to be taught, (ii) determining a criteria, for example, difficulty level, interest, sequence, to select the subject matter for particular population, and (iii) planning and implementing appropriate methods of instruction to ensure mastery of the subject matter selected. But this formula has certain limitations because the goal of education or even teaching of a particular subject is not restricted only to develop a mastery of the subject matter in the learner. It is further de-emphasized due to the recent movement of child centred curriculum. This type of criticism led curriculum experts to evolve a paradigm which can define curriculum in a better way. Tyler's model which identifies the paradigm as dimensions to define curriculum is simplest and quite acceptable for the planning and development of science curriculum in spite of some severe criticism. According to him the following four questions must be answered in developing any curriculum.

1. What educational purposes should the school seek to attain?
2. What educational experiences can be provided that are likely to attain these purposes?
3. How can these educational experiences be effectively organized?
4. How can we determine whether these purposes are being attained?

Later on Taba (1962) extended Tyler's four questions into step-wise interpretation in seven phases. These are (i) diagnosis of needs,

(ii) formulation of objectives, (iii) selection of content, (iv) organization of content, (v) selection of learning experiences, (vi) organization of learning experiences, and (vii) determination of what and how to evaluate. Further Dave and Dave (1991) have delineated four components of curriculum, namely (a) objectives defined in terms of expected learning outcomes, (b) content, (c) learning experiences (teaching and learning strategies) and (d) tools and techniques of evaluation. All these dimensions have led one to conclude that curriculum may be divided into the four related components' curriculum objectives, knowledge or content, learning experiences and curriculum evaluation. The present study tried to study three out of four components of the curriculum. The fourth component that is the evaluation aspect had not been studied which was the delimitation of the study.

The proposed research was based on the following assumptions:

1. There is need for a new curriculum which emphasizes science in a social setting, includes the application of science as central, stress local relevance in establishing a plan and utilize the many dimensions of science in planning students' experiences and teaching approaches (Yager and Avi, 1986). Throughout the 1980s science, technology and society programmes were developed in many countries. The focus was on societal issues and topics with a strong science and technology dimension. Later attempts were made by many science educationists like Aikenhead (1986), Bybee (1987), Layton (1988) to provide a conceptual framework for STS courses in developed countries but no such major attempt has been made in developing countries. There has been a constant need to reconstruct the science syllabus particularly in developing countries like India on the lines of relating it more effectively to contemporary personal, societal and national needs.

   The National Policy of Education (1986) also emphasized the need to create consciousness of the environment and to design science education programmes in such a way as to enable the learners to acquire problem solving and decision making skills and to discuss the relationship of science with health, agriculture, industry and other aspects of daily life.

2. There has been a constant need for content analysis of textbooks as it is considered to be an important teaching aid in conveying messages to the students about various aspects of the nature of science (Bhat, 1984; Brahaderswaran, 1986; Sarangapani, 1990). Some researchers even reported that the status of school science could be summarized with one word 'textbooks'. Ninety per cent of all science teachers used their textbooks in excess of 90 per cent of the time (Harms & Yager 1981; Stake & Easley, 1978; Weiss, 1978). Further, students expected the textbook to be used as the source of nearly all the information and as the framework from which all science was to be experienced. Even parents accepted the textbook as central and expressed concern if textbooks were not issued and used for assignments. Apparently textbooks provide an accurate record of what constitutes science study for nearly all students.
3. It should be ensured that STS themes are being included in the textbooks and more importantly, they should be included in the content in a meaningful manner. The most straightforward step to ensure this was to look through the existing content of the textbooks being taught and to identify the content where gaps existed and to ascertain how these could be filled. The auditing of the content can help in refocusing attention on the possibility of imparting STS themes either by modifying the presentation through new activities which will provide necessary experiences or by evolving a new strategy to transact them.
4. Mere theoretical formulation of STS themes and its incorporation in the content is not sufficient to remedify the lacunas existing in the curriculum. It can prove to be an engine of a successful revolution in education when the reformed content is field-tested in the real classroom situation. Thus, it was assumed that there was need to develop modified lessons for each class VI, VII and VIII in accordance with STS themes.
5. Simply teaching the content to the students does not ensure that students have perceived the concept and interpreted it the way researcher wanted. It is very essential to judge the

effectiveness of the modified lessons in terms of valid criterion variables.

Objectives of the Study

The Study was undertaken to—

1. Identify and validate the STS themes in science education at upper primary stage that suits Indian context.
2. Analyze the content of the Haryana State science textbook of classes VI, VII and VIII in terms of the weightage assigned to various validated STS themes in science education.
3. Identify the sub-units in the lessons where there is possibility of imparting STS themes either by modifying the presentation or by evolving a new strategy to transact them.
4. Suggest and exemplify the incorporation of STS input in textbooks of classes VI, VII and VIII.
5. Try out the modified lessons of class VII in one of the schools of Haryana and evaluate the effectiveness of these modified lessons in terms of achievement and understanding of science, technology and society.

Research Questions

1. What do Science, Technology and Society mean?
2. What are the STS themes relevant in the Indian context/society?
3. Do the science textbooks of upper primary level cover the STS themes?
4. Is there scope for integrating the STS themes in the present content without increasing the load by an extra subject?
5. Is the prepared exemplar lesson feasible in the real classroom situation?
6. Does the teaching of Science, Technology and Society concepts help in increasing the achievement and understanding of Science, Technology and Society?

Hypothesis

1. There is no significant difference between the experimental and the control groups with regard to the Understanding of

Science, Technology and Society when differential effect of intelligence and pretest was adjusted by ANCOVA.

2. There is no significant difference between the experimental and the control groups with regard to the achievement in science when differential effect of intelligence and pre-test was adjusted by ANCOVA.

Defining the Variables

The variables considered for the study fall under the following categories:

Independent Variable

Science, Technology and Society (STS) inputs are the content with STS themes and transactional strategies to be integrated in the already existing science content.

STS theme is presentation of scientific knowledge, skills and understanding in a personal/social context which encompasses the domains namely, (i) Acquisition of knowledge which covers all aspects of knowledge related to conventional content of science, science and technology taken together and interactions of science, technology and society. (ii) Skill and processes which cover aspects that can be used by students in the variety of situations and with which they explore and experience as to how technology interacts and technological changes impact directly on their lives. (iii) Ethics, values and attitudes domain embodies human feelings, values, ethics, decision-making skills and positive attitudes towards science, science classes, science teachers and the interactions of science, technology and society. (iv) Application domain focuses on technologies that are applications of science wherein students can relate their study to their daily life and helps in resolving societal issues.

Dependent Variable

(i) *Understanding of Science, Technology and Society*—Understanding of STS means understanding the nature of all the three aspects independently as well as in totality. In the present context understanding of STS includes:

- Understanding about the scientific enterprise.
- Understanding about scientists.
- Understanding about the methods and aims of science.
- Understanding about science and technology.
- Understanding the influence of society on science/technology.
- Understanding the influence of science/technology on society.
- Understanding the influence of school science on society.
- Understanding the social construction of scientific knowledge.
- Understanding the social construction of technology.

(ii) *Post-test (achievement test)*—It was defined as an index of student's achievement in terms of knowledge, comprehension and application of the subject matter (content covered in exemplar lessons) taught in the class.

Extraneous Variable

Covariates

(i) *Pretest*—It was defined as the index of student's knowledge and comprehension before the teaching of the subject matter (exemplar lessons), before the treatment.

(ii) *Intelligence Test*—In the present study intelligence is measured in terms of General Mental Ability by a verbal group test developed by Jalota (1976).This test treats intelligence as a possession. The term 'general mental ability' is defined as the ability to perceive and understand relationships such as logical, special, verbal, numerical and recall of associated meanings. The test used in the present study is loaded with three factors: (i) Verbal, (ii) Numerical and (iii) Reasoning. Therefore, the combination of these three abilities is operationally defined as 'general mental ability'.

Controlled Variable

(i) *Teacher Variable*—The same teacher, i.e. the investigator taught the experimental and the control group, thus the teacher variable was controlled.

(ii) *Socio-Economic Status*—This variable was controlled indirectly as students of both the groups were coming from similar socio-economic background.

(iii) *Time Effect*—The same amount of time was used to teach the experimental and the control group, thus the time effect was also controlled.

Sampling

Haryana, being a big agricultural state, requires innovations at technical and social levels. The new dimensions are to be introduced at the grassroots level, i.e. at the school level. The upper primary school level is the right stage to accelerate the pace of development as discussed earlier. The study was conducted in five phases. Sampling was required for the first and fifth phase of the study. In the first phase, sampling was done for experts and in the fifth phase for the students.

In the *first phase* for administering the rating scale to identify and validate the framework of STS themes in science education suitable in Indian context, purposive sampling was done to draw a sample of one hundred individuals belonging to three groups, namely (i) Subject teachers, (ii) Science activists and (iii) Science educationists. From the first group a sample of forty was drawn from a population of science teachers, teaching science at the upper primary level in the rural schools of Haryana. The population was restricted to only one district, i.e. Rewari. List of schools was obtained from the Office of the Directorate of Education and out of this list, random sampling was done to select twenty schools. Forty teachers teaching science at upper primary level were selected randomly from these twenty schools. For the second group, i.e. the science activist, thirty experts were selected. The further breakup of the sample was ten from the centre level activist group, i.e. National Council for Science and Technology Communication Network (NCSTC) for giving a nation-wide perspective, another group of ten experts was selected from the state level activists group, i.e. Haryana Vigyan Manch, which is the activists group working in Haryana and a group of ten experts was drawn from Eklavya in Madhya Pradesh as it is running an enquiry-based science teaching programme for middle school students. From the third group, i.e. science educationist, a sample of thirty was drawn. This group was further divided into subgroups consisting of experts from SCERT/DIET of Haryana, NCERT at a central level and other teacher training institutes at the state level of Haryana and Delhi, as Delhi is the state adjoining Haryana and being the capital, has many national and international level organizations which are actively involved in the educational innovations which are tried out at the teacher training institutes. A sample of ten each was drawn from each subgroup. A flowchart of the sampling of the first stage is given in Fig. 4.1.

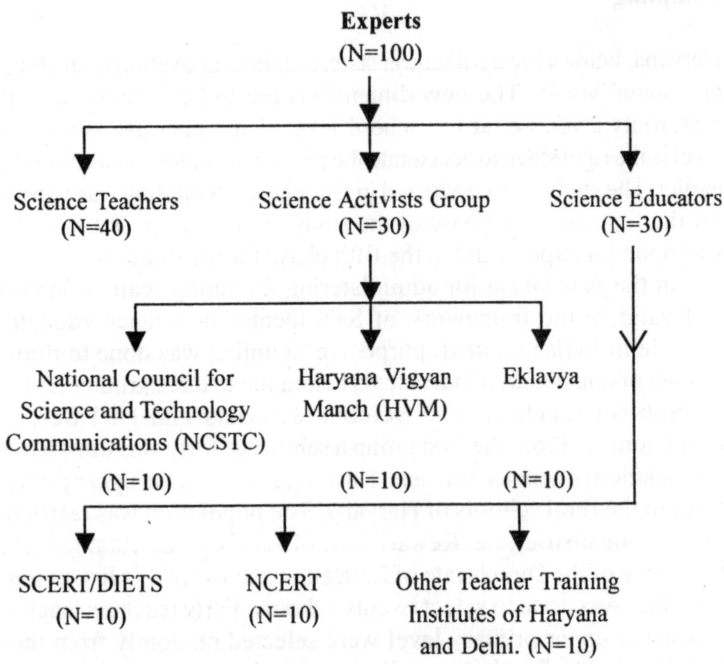

Fig. 4.1: Sampling for the Experts

To judge the effectiveness of the proposed exemplary lessons, an experiment was conducted at the *fifth* stage. For this purpose one of the schools was selected from the sample of twenty schools taken for the first stage of sampling. Two sections of class VII were selected for the experiment and from the two sections one section was randomly assigned as experimental group and the other was assigned as the control group. All the students of the two intact sections of class

VII constituted the sample of the present study. Initially the sample consisted of seventy-five students, i.e. thirty-six students of class VII-A and thirty-nine of class VII-B. As some of the students did not take all the tests, the resultant sample was reduced to seventy students, each group having thirty-five students. The two groups were as similar as the availability permitted, they were natural and highly comparable in respect of size, average age and past exposure.

Description of the Tools

After careful considerations following tools were selected or developed for collecting the data on the variables under study.

1. Rating Scale for Validation of STS Themes in Science Education for Upper Primary Science Curriculum in the Indian Context developed by the investigator.
2. Test on Understanding of Science, Technology and Society developed by the investigator on the basis of TOUS and VOSTS.
3. A Pre-test to measure the previous knowledge and an Achievement Test (Post-test) were developed by the investigator.
4. Group Test of General Mental Ability developed by S. Jalota.

Construction of the tools was done at two stages of the study. In the initial stage, to validate the STS themes framework in science education for upper primary level in the Indian context, and in the final stage of the study, to judge the effectiveness of the modified lessons, achievement tests (pre-test and post-test) and the test on understanding of science, technology and society. Brief descriptions of these tests are given below.

Rating Scale for Validation of STS Themes in Science Education at Upper Primary Stage in the Indian Context

The researcher developed STS themes framework in science education at upper primary level in the Indian context based on the related literature. It was constructed on the basis of the IOWA-STS dissemination model given by Yager and Tamir (1993), STS themes given by Bybee (1987), thinking constructively about STS by Dennis W. Cheek (1992), IOWA STS Project Student's Growth in Variety of Domain by Yager and Bruenkhoest (1987), NPE (1986), POA (1992), The National Curriculum for Elementary and Secondary Education: A Framework (1987), Syllabus for Upper Primary Stage (1987), Kakshawar Pathyakaram ki Masik Bat, Haryana Education Board (1993) and Curriculum for Middle Classes, Haryana (1987).

Five school teachers teaching science at upper primary stage in the schools of Haryana were asked to prepare a list of instructional objectives of science education at upper primary stage. The exhaustive list prepared by the investigator was compared with the instructional objectives of science given by the five science teachers. After comparison, those elements which were not present in the framework of the researchers but were present in the teachers' list were then included and the original framework was modified. A total of 78 aspects under the major domains of acquisition of knowledge, the skill areas and processes, ethics, values and attitudes and application area were included. The framework aspects were translated into Hindi by the researcher for better understanding by Haryana teachers. Two experts were consulted for the comprehensibility and the content relevance of the framework.

The tool was then given to a sample of hundred experts (details mentioned in the sampling). The experts were asked to respond on a three point scale including options like essential, important and not important with the numerical rating as three, two and one respectively. In addition to these items two open-ended questions were included to collect some qualitative data about the conceptual framework of STS themes. A data sheet was prepared for the numerical rating of the experts. The numerical ratings for each aspect under each domain were totalled and means were calculated. The aspects rated by the experts as not important, that is having a mean value less than 2.5, were discarded and aspects having mean value 2.5 or more were retained. A total of thirty-seven aspects were retained out of seventy-eight, on the basis of the experts' opinion. The details of the tool are given in Appendix B.

Test on Understanding of Science, Technology and Society (USTS)

This test was developed on the basis of the items of the Hindi version, Test on Understanding of Science (TOUS) originally developed by Klopfer, Shoresman, Carrier and Gies (1970) and adapted by Masih for his doctoral study and Views on Science, Technology and Society (VOSTS) developed by Glen S. Aikenhead, Alan G. Ryan and Reg W. Fleming (1989).

TOUS was developed at Learning Research and Development Center, University of Pittsburgh, USA. It consists of 36 multiple-choice items with four alternatives (one most appropriate answer and three distracters). The present form used was a translated Hindi version, translated and used by Masih (1990). The reliability of the present Hindi version of TOUS in terms of inter item consistency was calculated by Kuder-Richardson formula 20.The reliability coefficient was found to be 0.395. This value, although not very high, can be considered quite adequate.

VOSTS is an assessment instrument dealing with STS topics. VOSTS consists of a pool of hundred and fourteen multiple-choice items that address a broad range of STS topics. The main aspects covered under this conceptual scheme are science and technology, influence of society on science/technology, influence of science/ technology on society, influence of school science on society, characteristics of scientists, social construction of technology, social construction of scientific knowledge and nature of scientific knowledge. The items were developed for high school students' view on a broad range of STS topics. The validity of the process and of the final instrument, as stated by the developer of the tool, lies in the trust which subsequent researchers place in the process which has been used by the authors. From the test items given in VOSTS, only twenty-eight items were chosen by the researcher with the help of the five experts for the students of the upper primary stage, as the other items were difficult for the students of this stage to understand. The items were constructed in a simpler form and translated into Hindi for the comprehension of the students. During the translation work, the investigator was very alert and every possible precaution was taken so that the originality of the theme of each item remained intact. Whenever she felt any difficulty or confusion, she consulted the experts of both the languages (i.e. English and Hindi) as well as some science educators. All possible efforts were made to maintain the spirit and emphasis of each word of the text intact in the translation. When the investigator felt satisfied with her work, the initial draft was shown to the supervisor and necessary corrections and modifications were made. It was also sent to the five experts to see the appropriateness of translation and its utility in Indian situation.

The translated version of the test was modified in the light of the experts' comment.

The modified form was administered to a sample of twenty students of class VII of the government school in Rewari, Haryana for the pre-try out. This school was different from the school chosen for the main experiment. The entire process of administration of the tool was very informal. The students were free to ask any question regarding the test. The student's remarks and difficulties were noted immediately. The purpose of this pre try out administration was to collect students' reactions about the comprehensibility and clarity of the test in terms of language, gross deficiencies, ambiguity in statements, inadequacy in the direction to the examinees and to collect information about the appropriate time limit to complete the test. The items were modified again and then administered on three hundred students of three schools in Rewari. The twenty-eight items were then analyzed for their difficulty and discriminating values. The items with difficulty values ranging from twenty-five to seventy-five and discrimination values of 0.45 and above were retained and the rest were dropped. After item analysis, only fourteen items were retained which were added to the thirty-six items of the TOUS. The new test, that is test on Understanding of Science, Technology and Society (USTS) consisted of a total of fifty items, as given in Appendix C. Since reliability coefficient of this new test was not known, the reliability of the test was calculated by Kuder-Richardson formula 20. The value of the reliability coefficient was calculated by administering the test to a sample of one hundred students of class VII randomly selected from different sections of Government Girls Senior Secondary School, Rewari. The reliability coefficient was found to be 0.57.

Achievement Test (Pre-test and Post-test)

Both pre-test and post-test were constructed to administer on both the control and experimental groups before and after the experiment. The control and experimental groups consisted of the students of two sections of class VII randomly assigned as control and experimental. Pre-test was constructed to check the pervious knowledge of the two groups, before exposing them to the treatment. The pre-test was based on the lessons before the main lesson taught to them, i.e. population. A total of fifty multiple choice items were prepared from the lessons before the lesson of population in the class VII science textbook.

The post-test was constructed to check the achievement of the two groups after the treatment. This test was based on the lessons Population and Pollution. Care was taken while constructing the test items of this test as it had to be comparable in terms of the content taught to both the groups. A total of fifty multiple choice items were prepared for the test.

The items for both the tests were limited to knowledge, comprehension and application categories of the educational objectives of cognitive domain. These tests were sent to the five experts to review the test items in terms of their appropriateness of content, flaw in the language (if any) and their technical appropriateness. Based on the experts' suggestion and remarks the items were modified. The two tests were pre-tried out on a sample of twenty students of class VII of the government school in Rewari. The test was administered under quite normal conditions in the environment of a class. To make rapport with the children, the purpose of the test was briefly discussed with them and thereafter the test booklets were distributed. The investigator read out the instructions and the student was allowed to ask any question about the test. The students were also instructed to write the time of starting the test; for the convenience of the students an alarm clock was kept in the class. The investigator noted down the difficulties faced by the students. The students were asked to write down the time of completion of their test. The items were modified again and then administered on three hundred students of three schools in Rewari. The performance of each item by the students was then scored with the help of a scoring key, each correct response was assigned one mark while the incorrect one was assigned zero. Answer sheets were arranged in descending order of scores for item analysis.

The difficulty value and the discriminating value of each item were calculated. The difficulty value (facility value) of a test item is indicated by the percentage of students who get the item right. So the difficulty (facility) value is inversely related to difficulty of the item under direct measure of easiness while the discrimination value of an item is indicated by the difference between good and poor students in proportion of correct response. Besides the concentration on difficulty value and discrimination value of individual item, the proper representation of items related to three different categories of cognitive domain was also taken into consideration while selecting items for final tests. The minimum and maximum value of the difficulty and discrimination indexes of selected items of the two tests are given in Table 4.1.

Table 4.1: Minimum and Maximum Difficulty Value and Discrimination Value of Selected Items

Name of the Test	*Difficulty Value*		*Discrimination Value*	
	Maximum	*Minimum*	*Maximum*	*Minimum*
Pre-test	55	28	0.65	0.47
Post-test	64	30	0.69	0.42

The distribution of items included in final test, according to the educational objectives of cognitive domain are given in Table 4.2.

Table 4.2: Distribution of Items Included in the Final Test According to the Educational Objectives

Name of the Test	*Educational Objective*			
	Knowledge	*Comprehension*	*Application*	*Total*
Pre-test	8	14	8	30
Post-test	6	12	12	30

As shown in Table 4.2, in the Pre-test and the Post-test, the maximum number of items is in comprehension category of cognitive domain followed by knowledge and application categories.

After the item analysis of both the tests only thirty items of each were retained from the fifty items prepared originally, which were used for the final study (as given in Appendix D). The reliability coefficient calculated by Kuder-Richardson formula-20 was found to be 0.74 for the pre-test and 0.61 for the post-test.

Group Test of General Mental Ability

Group Test of General Mental Ability (1976) is a revised form of its earlier version, developed in 1951-52. It covers the elements of (i) vocabulary similar, (ii) vocabulary opposites, (iii) classification, (iv) best number, (v) analogies, (vi) reasoning, (vii) number series. It consists of three sub-scales, viz., (a) Verbal Ability Scale, (b) Reasoning Ability Scale and (c) Numerical Ability Scale. The reliability and validity of the test is very high as tested by many researchers in their studies (given in Appendix E).

Procedure

The study was carried out in the five phases as shown in the phased diagram given in Fig. 4.2.

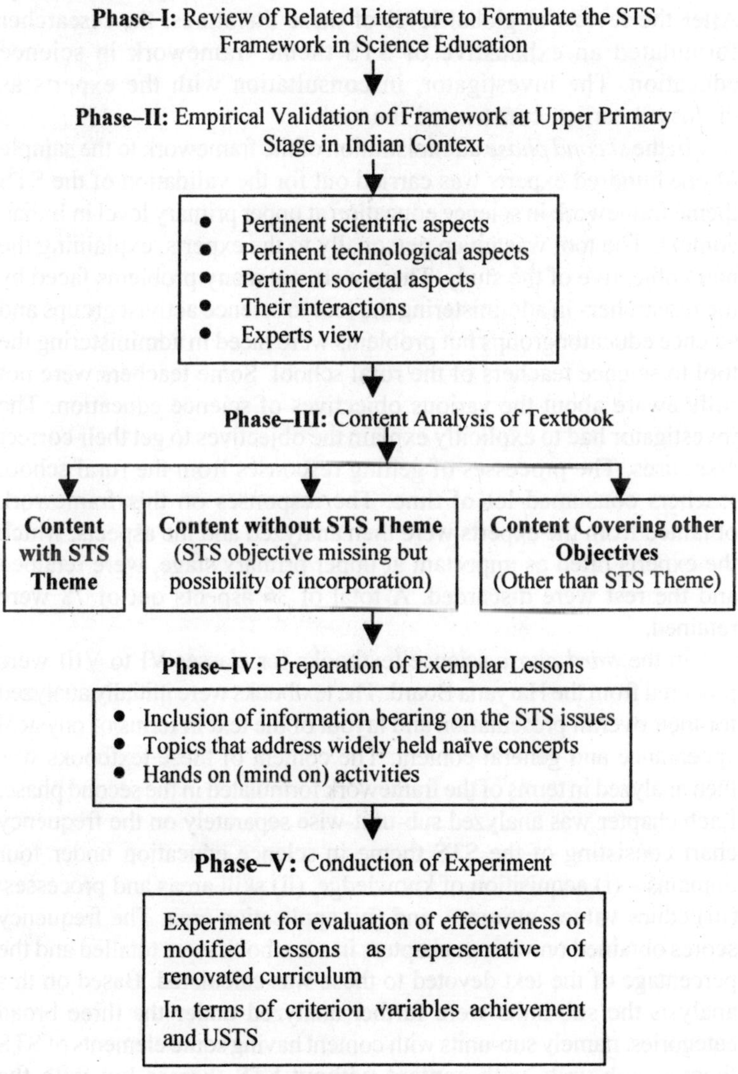

Fig. 4.2: Procedure of the Study

In the *first phase*, visiting of various libraries and institutions was done initially to carry out a thorough review of the related literature. Some literature was surfed through Internet and some was procured from abroad by post. Various books, journals and reference materials were also consulted on science, technology and society. After the review at global level of these literatures, the researcher formulated an exhaustive of STS theme framework in science education. The investigator, in consultation with the experts as discussed earlier, constructed the tool.

In the *second phase* administration of the framework to the sample of one hundred experts was carried out for the validation of the STS theme framework in science education at upper primary level in Indian context. The tool was given personally to the experts, explaining the main objective of the study. There were not many problems faced by the researchers in administering the tool to science activist groups and science educator groups but problems were faced in administering the tool to science teachers of the rural school. Some teachers were not fully aware about the various objectives of science education. The investigator had to explicitly explain the objectives to get their correct responses. The processes of getting responses from the rural school teachers consumed lot of time. The responses on this framework obtained from the experts were then analyzed and the aspects, which the experts rated as important at upper primary stage, were retained and the rest were discarded. A total of 36 aspects out of 78 were retained.

In the *third phase* science textbooks for classes VI to VIII were procured from the Haryana Board. The textbooks were initially analyzed for their overall presentation and layout of the text in terms of physical appearance and general content. The content of these textbooks was then analyzed in terms of the framework formulated in the second phase. Each chapter was analyzed sub-unit-wise separately on the frequency chart consisting of the STS theme in science education under four domains—(i) acquisition of knowledge, (ii) skill areas and processes, (iii) ethics values, attitudes, and (iv) application area. The frequency scores obtained on various chapters in a textbook were totalled and the percentage of the text devoted to these was calculated. Based on this analysis the sub-units were further analyzed under the three broad categories, namely sub-units with content having some elements of STS themes, sub-units with content without STS themes but with the possibility of incorporation and sub-units with content without STS themes but covering other objectives of science education other than related to STS.

In the *fourth phase* two lessons from each classes VI-VIII textbook having sub-units with the maximum possibility of incorporation of STS themes were identified for inclusion of information bearing on the STS issues. The sub-units covering other objectives of science education other than related to STS were not touched and were kept as such. The lessons which had maximum sub-units with the possibility of STS incorporation were modified, which were 'Measurement 'and 'Water' of class VI, 'Population 'and 'Pollution' of class VII and 'Agriculture' and 'Refraction' of class VIII. After the content analysis of the textbooks and modification of the six chapters, tools were constructed for the experiment and tried out as already discussed in the construction of the tool.

In the final phase i.e. the *fifth phase*, the modified lessons of class VII were tested in real classroom situation. Since the true experimental design requires random selection of the subjects, this requirement was difficult to be realized in a real classroom situation, the 'Non-Equivalent Control Group Design' (Campbell and Stanley 1986) was chosen for the study. This is the most wide spread experimental design which is used in educational research. It involves an experimental group and a control group, both given pre-test and post-test treatment, but in which the control group and the experimental group do not have pre-experimental sampling equivalence. Rather the groups constitute naturally assembled collectives such as classrooms, as similar as the availability permits. The treatment given to one group or the other is assumed to be random and under the experimenter's control. This design maximizes the internal validity by controlling the main effect of history, maturation, testing and instrumentation (variables).

The difference for the experimental group between the pre-test and post-test (if greater than the control group) cannot be explained by main effect of these variables such as would be found effecting both the experimental and control group. The quasi-experimental design consisted of one experimental group and one control group. The experimental group was taught the modified lessons with more STS orientation by relating the content to STS issues and the control group was taught the conventional content. The teaching strategies while teaching both the groups were kept more or less alike to judge the true effectiveness of the STS oriented content on the achievement and understanding of science, technology and society.

The experiment was conducted in Government High School, Bharawas. The investigator obtained prior permission of the principal of the school and a schedule for conduction of the experiment was fixed. The investigator herself taught the experimental and the control groups. Twenty lesson plans were prepared in total for the chapters Population and Pollution, prior to the teaching. Ten lesson plans each were prepared for the control group and the experimental group. Four lessons were prepared for the chapter on Population and six lesson plans were prepared for the chapter on Pollution. A sample of each lesson plan (experimental and control) is given in Appendix F.

Lesson plans prepared for the control group were based on the content given in the textbook. The content was taken intact and only those activities were transacted which were given in the textbook. It would be appropriate to mention that there were as such no activities given in the main lessons on population and pollution for enhancing the comprehensibility of the concepts. While for the experimental group lesson plans were prepared of the same content but with STS orientation either by exemplifying them with day to day examples or by incorporating activities, i.e. performing role play on small and big family to introduce the concept of population or by telling the story of a farmer as to how his acres of land went on getting divided with each generation because of increase in the number of family members. The transaction of two kinds of lesson plan took almost equal time as in the traditional lesson plan students were given class assignments where they had to give examples from their life or different paragraphs were assigned to different students which they read aloud and the researcher explained.

Pre-test and the General Mental Ability Test were administered to the students before teaching was started in the two groups. Post-test was given first to the students, instructions were written on blackboard and were explained verbally. In all, thirty multiple-choice items were given in the test and the students were asked to complete it within 35 minutes. When the students completed this test a break of ten minutes was given and after the break Group Test of General Mental Ability was administered. The students were told in advance that they had to finish this test strictly in twenty minutes. They were told that the test contained one hundred questions. Every question is followed by a number of answers and students had to mark the number of the correct answer in the separate answer sheet provided along with the test booklet. The instruction was readout by the investigator, then all the fourteen items given as examples were solved by the investigator with the help of students and only then did the investigator ask the students to 'start'. The time of starting was noted down and exactly after twenty minutes they were asked to stop writing and the test booklets and answer sheets were collected back. Both the tests were given simultaneously to both the groups.

The whole experiment was conducted in two weeks. Two lesson plans were taught in one day. On day one, the experimental group was taught first and then the control group. However, on the next day the sequence was reversed. Special care was taken while teaching that no group should be exposed to more than forty minutes of instructional treatment per day. The teaching of two groups was done in one day separated by at least one hour. This was done to minimize the element of biased-conditioning in the researcher and to prevent the carry over effects of one teaching strategy into the other by the researcher as a teacher. Another purpose was to avoid undue cognitive strain on the pupils due to prolonged learning environment conditions.

On the completion of the teaching to both the experimental and the control groups the post-test and USTS were administered. Post-test was administered first. The students were asked to give personal information like their name, name of the school, class, date etc. on the space provided in the test sheet. The investigator gave instructions by writing on the blackboard and explaining verbally. There were in total thirty multiple-choice items. The time taken by students in both the groups was thirty-five minutes. After the completion of the post-test, a break of ten minutes was given and the students told to assemble again for the next test. When the break was over Test on Understanding of Science, Technology and Society was administered. The instructions were again written on the blackboard and even read aloud. Since time limit was not specified in the original form of this test, the students were instructed to finish the test as soon as practicable. There were in total fifty multiple-choice items and the students had to choose one correct response out of the alternatives. The time taken in general by both the groups was sixty to sixty five minutes. Before permitting the students to leave the room, answer sheets were carefully scanned to see whether all the items were answered or not. If any blank was left the students were asked to complete the same before leaving the class. Both the experimental and control groups were given both the test simultaneously to avoid any discussion of questions among the students of the two groups which would have affected the results.

After the completion of the experiment, the students, principal and teachers of the respective sections were thanked for their cooperation.

Scoring

The answer sheets of the entire four tests were scored manually by using scoring keys prepared for the respective tests. In the post-test and pre-test each item correctly responded was assigned one mark while zero mark was given to an item incorrectly responded. The Test on Understanding of Science, Technology and Society consisted of fifty multiple choice items with four alternatives each. There was one best answer for each question and each item was assigned one mark if the respondent chose the best answer while a zero mark was given to other responses. The scoring of answer sheets of Group General Mental Ability Test was done by using scoring stencil made from the scoring key according to the directions written in the manual of the test. The correct answer was given in the scoring stencil and it could be easily compared with the student's answer of the said item by putting the stencil over the solved answer sheet. Each item, if correctly responded, was assigned one mark. For convenience the number of wrong responses and unattempted items were marked with red pencil and subtracted from the total number of items; the residual gave the marks obtained by an individual in the test. The total score of an individual in this test may range from 1 to 100. Higher score means higher mental ability.

Statistical Treatment

ANCOVA was used to analyze the quantitative data and content analysis was done to analyze the textbooks. The detailed description of data, their tabulation and analysis are presented in future chapters.

Delimitations

Keeping in view the background of the problem and research design, along with the constraints of time and resources, the present study has been delimited in following ways:

1. The renovation of the science curriculum was restricted to STS framework, its incorporation in the textbooks and its transaction.

2. Areawise study was confined to only one district of Haryana, i.e. Rewari.
3. The study was limited to upper primary stage (classes VI–VIII) and to one subject, science.
4. Only two lessons each from each class VI, VII and VIII textbooks were modified in the light of STS orientation.
5. The experimental as well as control treatments, in the form of instructions, were given by the investigator herself.
6. The experiment was conducted only in one school of Haryana and both experimental and control groups were selected from the same school.
7. The evaluation techniques aspect of the curriculum was not studied in the present study.

5

STS Themes in the Indian Context and their Validation: Responses of Experts

Introduction

Science education as a discipline is concerned with the interface between science and society. It must respond to social changes, and various societal issues should be reflected in school science programmes. The goal of science education must be reformulated to include the personal and social dimensions that have been ignored for over two decades. There has been professional debate centring on the domain of science education and stimulated by the issue of locating science-technology-society topics within the discipline. Good *et al.* (1985); Yager (1985); Hofstein and Yager (1982) advocated debate about the definition of science education, particularly in reference to the inclusion of STS themes. According to them, science education must respond to the social changes and various societal issues should be reflected in the school science programme.

The 1985 National Association for Research in Science Teaching (NARST, Indiana) meeting included a symposium on "Science Education: Definitions and Implications for Research" where emphasis was laid on the emerging STS theme as central to science education. The position presented was that traditional science focuses almost exclusively on developing conventional knowledge and gives little allowance for goals related to the STS theme in science education.

Two important points of elaboration and clarification were inferred from various debates and presentations, one that STS ought to be included in science education programme as science education is a social institution and second, the use of the preposition is used in the phrase *STS theme in science education*, which implies that STS themes are only a part of science education. It is central to science education but at the same time it should not exclude the other goals of science. The theme of science-technology-society has been on the national agenda in science education abroad (NSTA, 1982; Bybee, 1986; Yager, Hofstein and Luneta, 1981; Hofstein and Yager, 1982; Hurd, 1984). But none such work has been carried out in India. There is an inbuilt element of science and technology in the objectives of science education at the upper primary level but it is not explicitly mentioned and is generally overlooked by teachers and curriculum planners. In order to have personal and social goals in science education it is essential to review the objectives of science education.

STS Theme Framework in Science Education

A STS theme framework in science education clarifies the understanding of how to proceed about STS issues. It fulfils several important developmental and pedagogical criteria for the translation of concepts, processes and understanding into curriculum programmes and instructional practices. Thus, there is a need for STS theme framework in science education, which includes primary areas in which objectives, which are formulated, are based on scientific and technological knowledge, processes and personal/social application. Keeping this in mind an exhaustive list of STS theme objectives within science education was made by the researcher on the basis of the concerned literature, namely, Iowa-STS dissemination model given by R.E. Yager and P. Tamir (1993), STS themes given by R.W. Bybee (1987), thinking constructively about STS by Dennis W. Cheek (1992) Iowa STS Project *Student's growth in variety of domains* by R.E. Yager and B.J. Bruenkhoest (1987), NPE (1986), POA (1992). The National Curriculum for Elementary and Secondary Education: A Framework (1988), Syllabus for Upper Primary Stage (1987), Kakshawar Pathyakaram ki Masik Bat, Haryana Education Board (1988) and Curriculum for Middle Classes, Haryana Education Board (1987).

Five school teachers teaching science at upper primary stage in the schools of Haryana were asked to prepare a list of instructional objectives of science education at upper primary stage. The exhaustive list prepared by the researcher was compared and matched with the list of instructional objectives given by the five school teachers teaching at the upper primary level. After comparison, those elements which were not present in the STS framework of the researcher but were present in the teachers' list, were included. The STS theme in science education framework originally prepared in English was also translated in Hindi by the researcher with the help of the experts. It consisted of the exhaustive list of all the possible aspects in each domain.

Many science educators believe that a "strong emphasis on science-technology-society topics would increase the relevance of material and decrease the compartmentalization found too often in current courses, but traditional goals of science education should not be eliminated to make way for the STS theme" (Brincherhoff and Yager, 1986). Keeping this in mind the STS framework had a blend of traditional goals and themes of STS in science education.

The framework consisted of four major domains, namely acquisition of knowledge, skill areas and processes, ethics, values and attitudes, and application area. The domain acquisition of knowledge was further subdivided into three sub domains that is, concepts related to conventional content of science, concepts related to science and technology and concepts related to science, technology and society. The detail of the STS framework is given in Appendix B.

In the domain acquisition of knowledge under the sub-heading concepts related to conventional content of science, there were six aspects, covering all the aspects from facts to existing explanations for these facts. Under concepts related to science and technology, there were ten aspects covering all the aspects from knowledge of technological system and subsystem to technological pattern and processes to its equilibrium and sustainability. Under concepts related to science, technology and society, there were fourteen aspects covering aspects from understanding the interactions of science, technology and society to global awareness, to being a critic of proposed solutions in STS issues.

In the domain skill areas and processes, there were twenty-six aspects ranging from nature of scientific inquiry to communicating skills regarding STS issues to developing creativity skills. In the domain ethics, values and attitudes, there were twelve aspects ranging from aesthetic appreciation of the world, science and technology to valuing democracy and democratic process to making decisions about social environmental issues and exploring arguments on various sides of an issue.

Table 5.1: STS Themes Framework in Science Education

Acquisition of Knowledge		
Concepts Related to Conventional Content of Science	*Concepts Related to Science and Technology*	*Concepts Related to STS*
• Facts • Information • Concepts • Laws	• Interaction & Change • Growth & Cycle • Patterns & Processes • Probability & Prediction	• Describe Interaction of ST&S • Understands limits of S&T • Distinguish Personal Opinion from Scientific Evidence • Understand Application of Technology
Skills and Processes		
• Questioning & Searching • Critically Thinking Skills • Observing, using space-time relationship • Use of Technology in Practical Situation • Comparing and Conserving • Suggesting possible causes of a given situation		• Classifying, grouping organizations • Combining objects and ideas in new ways • Interpreting Data • Predicting Possible consequences • Producing Mental Images
Ethics Values and Attitudes		
• Value of democratic process • Identifying values bearing on any given STS issues like local issues, public policies, global problems • Developing positive attitude towards science in general, science in school and science teacher • Developing positive attitude towards oneself • Expressing personal feeling in a constructive way • Developing sensitivity to and respect for the feeling of other people • Making decisions about social environment issues and exploring arguments on various sides of an issue		
Application		
• Seeing instances of scientific concepts in everyday life experiences • Using scientific process in solving problems in everyday life • Extended school experience beyond classroom • Understanding Science and Technology principles involved in household devices • Making decisions related to personal health, nutrition and lifestyle based on knowledge of Scientific Concepts • Handling of apparatus • Development of low cost materials		

In the domain application area there were eleven aspects ranging from seeing instances of scientific concepts in everyday experiences to making decisions related to personal health, nutrition and lifestyle based on knowledge of scientific concepts to conducting research and taking action on STS issues. In total there were seventy-eight aspects under the different domains.

The framework on STS themes in science was given to a sample of hundred experts. These experts were asked to rate the aspects on a three point rating scale, namely essential, important and not important at upper primary level in the Indian context. The responses received on the rating scale were tabulated and then analyzed. The responses were assigned a numerical rating of 3, 2, 1. Individual means were calculated for each category in the domains. The category which had a rating of 2.5 and above, was retained and the rest were discarded. Out of the seventy-eight aspects kept at the initial stage only thirty-seven were retained.

Aspects that were discarded were existing explanation, system and subsystems, organization and identity, hierarchy and diversity, adaptation and limitation, equilibrium and sustainability and many others in the various domains which, probably the experts did not think were important at the upper primary stage in India or for some other reasons did not rate them as essential. The thirty-seven aspects validated were under four different domains as given in Table 5.1.

Acquisition of Knowledge

The first domain, the acquisition of knowledge, covers all aspects of knowledge acquisition, which have been very intricately woven into it. This domain focuses on the more conventional content of science, content related to science and technology, content related to science, technology and society. Part of any science instruction involves student learning of the concepts central to science. Students' understanding of current explanations of objects and events in the universe is important, particularly when students can also use the concepts learned. Knowledge continues to expand rapidly with the changing technology. Students should know the interactions between science, technology and society and the changes occurring due to these interactions. There were twelve aspects validated under this domain with a break-up of: four under the sub-heading concepts related to conventional content of science, four related to science and technology and another four related to science, technology and society (Facts, information, concepts, laws, interaction and change, growth and cycle, patterns and processes, probability and prediction, describe interaction of science, technology and society, understand limits of science and technology, distinguish personal opinion from scientific evidence and understand application of technology).

Skills and Processes

The second domain, skills and processes, displays science as a narrative of inquiry instead of a rhetoric conclusions. It is concerned with how scientists think and how their work represents another dimension of science. There are definable processes that characterize human actions that result in new knowledge about the universe and the scientific and technological developments. These skills and processes can have applicability for all learners. To know them means being able to use them in a wide variety of situations; students will explore and experience as to how technology interacts and technological changes impact directly on their lives. This domain includes aspects which can be acquired or constructed when pertinent STS issues are taught. Eleven aspects were validated under this domain (Questioning and searching, observing using space-time relationship, classifying grouping and organizing, interpreting data, comparing and conserving, critically thinking skills, use of technology in practical situations, producing mental images, combining ideas in new ways, predicting possible consequences, suggesting possible causes of a given situation).

Ethics, Values and Attitudes

The third domain consists of ethics, values and attitudes that embody the values which are inherent in chosen STS issues. It is generally accepted that all learning is enhanced when students have positive attitudes about the subject. They are more motivated, more involved with meeting the course objectives, and are more likely to realize instructional goals. Negative attitude discourage many and tend to make programme goals trivial, even for those who wish to generally excel in school. The focus of STS upon current issues and students' questions and interests makes the attitude domain particularly important. In these times of increasingly complex social and political institutions, environmental and energy problems, and general worry about the future, scientific content, processes, and even attention to imagination are not sufficient parameters of a science programme. Human feelings, values, ethics and decision making skills need to be addressed. Exemplary science programmes must be concerned with the generation of more positive student attitudes about science, science classes, the usefulness of science studies, science teachers and science courses. Seven aspects were validated under this domain (Values of democracy and democratic process, identifying values bearing on any given STS issue like local issues, public policies, global problems, developing positive attitude towards science in general, science in school and science teacher, developing positive attitude towards oneself, expressing personal feeling in a constructive way, developing sensitivity to and respect for the feeling of other people, making decisions about social environment issues and exploring arguments on various sides of an issue).

Application

It is pointless to have a programme that does not include a substantial amount of information, skills and attitudes that can be transferred and used in students' everyday lives. It is inappropriate to divorce academic science from technology. Students need to become sensitized to those experiences they encounter that reflect ideas they have learnt in school science. To be meaningful, the material in science courses should be related to real life situations. It should be focussing on technologies that are applications of science, such applications also provide connections and serve as a means of enhancing creativity and attitude. When these are enhanced more persons will see the value of processes and concepts and will advance to these commonly emphasized domains because they realize their relevance, rather than accepting teacher's assertions that they will be useful in the future. The fourth domain consists of application which covers aspects wherein the students can relate their study to their daily life and become involved in resolving societal issues. Seven aspects were validated under this domain (Seeing instances of scientific concepts in everyday life experiences, understanding science and technology principle involved in house hold devices, using scientific process in solving problems in everyday life, making decisions related to personal health, nutrition and lifestyle based on knowledge of scientific concepts, extended school experience beyond classroom, handling of apparatus and development of low cost materials).

The framework fulfils several important developmental and pedagogical criteria for the translation of concepts, processes and understanding into curriculum programmes and instructional practices. Progression down the columns in the framework leads from personal to cultural, information gathering to decision making and local to global perspectives.

Overall review of the framework suggests an orientation towards the personal and social purposes of science education, while examination of the separate aspects in the domains identifies specific objectives in the area of knowledge, ethics, values, skill areas and processes and application areas in STS interactions. The next stage is the identification of the most effective means of incorporating this framework of STS themes in science education into the existing science programmes, instructional materials, in service programmes, strategies and textbooks.

6

Coverage of STS Themes in Science Textbooks: An Overview

Introduction

Once a new science policy is evolved and goals identified, the first step taken immediately is the development of a curricular framework. With the goals in view, the textbook is produced. Although in educational terms, "curriculum" is the sum total of the activities that a student undergoes in school, in practice it only means the textbook. Most teachers depend almost exclusively on textbooks to define their courses and to provide the activities for use in classrooms, and teaching strategies to deal with them. Teachers see a good textbook as one that makes life easy for the teacher users and keeps the students engaged.

Some studies in the eighties revealed that ninety per cent of all science teachers use a textbook (page by page) in excess of ninety per cent of the time. For most students science becomes what is printed in textbooks and what is included on associated worksheets and in verification type laboratories. (Helgeson, Blosser & Howe, 1977; Stake & Easley, 1978; Weiss, 1978). According to Yager, (1985) the status of science education can be summarized in a single word—textbooks. The "Project Synthesis" (Harms and Yager, 1981) combined the findings of the three NSF studies with that of the National Assessment of Education Progress, USA, according to which, ninety percent of all science teachers use a text book ninety five per cent of the time, hence the textbooks becomes the course outline, framework, the parameters of the students' experience, testing and a world view of science. Nearly all science teachers emphasize (ninety per cent) goals of school science that are directed only to preparing students for further formal study of science.

The National Science Foundation (NSF), USA funded three separate studies to look into the status of science education in the United States (Hegelson *et al.*, 1977; Weiss, 1978; Stake & Easely, 1978). One of the major findings of the NSF studies was the enormous dependence of science teachers on the textbooks. The following observation of the NSF reports bear testimony to the importance given to the textbooks:

- Textbook determines the order, the example and the application of science topics that students experience.
- The science curriculum is the content in science textbooks.
- Teachers seek safety in maintaining close alliance with the textbook.
- Teacher's powers seem to rest with the information in a textbook.
- What a given student knows, does, thinks regarding science can be approached by reviewing two or three commonly used textbooks in various disciplines of science and/or levels of instructions.

Although such significant studies have not been carried out in India, it is a well-established fact that textbooks play a central role in our school education. One of the findings, of a study of the system of science education in India, was that textbooks were the only material through which curriculum guidelines percolated via the practicing schools (Menon, 1986).

The National Policy on Education (1986) emphasizes that science education upto secondary level may be imparted as one area of human endeavor rather than compartmentalized sub-disciplines. The National Curriculum for Elementary School—A framework developed by NCERT in 1985 and revised in 1988, provides a concrete structure to develop a relevant science curriculum reflecting the goals of NPE 1986.

In India, textbooks are prepared keeping in view the goals of NPE and the structure given by the National Curriculum Framework (1988). The main emphasis is to give concept learning to students rather than passing on mere information. The content and activities included in the textbooks accordingly should aim at developing well defined abilities in cognitive, affective and psychomotor domains such as spirit of inquiry, creativity, objectivity, courage to question and aesthetic sensibility. The messages contained in a text influence how the users (primarily the students, but also teachers) perceive the subject. Thought should be given as to how the textbooks represent science. This would indicate the probable ideas and notions about science that the users receive.

In today's world science, technology and their products have become a part of human life. Therefore, how the text conveys the interactions of science, technology and society would be critical in communicating the importance of their interaction and teach the responsible use of the products to the students. It is, therefore, imperative that the science textbooks should be a fine blend of science, technology and society so as to enable the learner to acquire problem solving and decision making skills and to discover the relationship of science with health, agriculture, industry and other aspects of daily life. The textbook should not only make a child scientifically literate but also technological literate; along with making students literate, it should inculcate universal human value as very rightly put up in the National Curriculum Framework for School Education—A Discussion Document prepared by NCERT (2000)—"Textbooks have to be written carefully and also reviewed frequently to assess what values they are communicating."

It was thus thought to analyze the science textbooks of upper primary stage prepared by Haryana Education Department. The science textbooks of classes VI, VII and VIII were firstly analyzed in terms of their general presentation layout and then in terms of their utility from the point of view of science, technology and society. The study does not go into the basics of the curriculum or the syllabus for the classes but explores as to how to make the existing content more STS oriented wherever needed without altering the main concepts, and to be taught within the same time frame without exerting an extra burden either on teachers or on students at the same time making teaching learning more child centered and skill oriented.

General Presentation and Layout of the Textbook

The textbooks were first analyzed on their general presentation and layout. The physical get up of the science textbooks of classes VI, VII and VIII were found as drab, unattractive and not very catchy. The binding of the book was not satisfactory; if frequently used by the student it was bound to be loosen.

The paper used for printing was of poor quality. The colour of the paper was neither white nor off-white, the texture and quality of the paper was rough and inferior. The printing ink used and the printing output was also poor. Many of the pages like 10, 11, 27, 70, 87, 97, 126, 127, 130, 131 of class VI textbook, pages 7, 10, 25, 31, 54, 79, 140, 149, 151, 152, 153 of classes VII textbook and pages 51, 53, 58, 66, 67, 70, 71, 79, 159, 177 of class VIII textbook have very poor readability as the printing was smudged and the words of the content were not clear. Let alone the children, even the researcher had difficulty in reading these pages.

Language of a book is an important tool for communication of ideas between the author and the reader. The language used should be simple and comprehensible. In some lessons the language of the books is simple and within the understanding levels of the children but in other lessons the language is tough to understand. Certain difficult Hindi words used in these books are, for example, words like संहति, आंकलन, द्रवणांक, गलनांक, क्वथनांक, पारभासक, संचित, संकुलन, प्रत्यास्थता, आसवत, उर्ध्वपालन, घूर्णन, उत्तोलक, आनतजल, उत्सर्जन used in Class VI textbook उप्पलावन, लम्बवत्, पाठांक, खेदार, अवशोषित, वितिकरण, सवंहन, आवेशन, प्रतिकर्षित used in Class VII textbook and words like अभिलम्ब, अनुप्रस्थ, अपेक्षाकृत, स्वचालित, सांकेतिक, युक्तियों, नगण्य used in class VIII textbook, which were difficult for the students to comprehend. The language of the books should be simplified as it consumes lot of time to explain the difficult terms and words. Very simple and small words should be used and as far as possible emphasis should be on the comprehension of the concept and not on the development of the vocabulary.

The content presented was less illustrative whereas science textbooks should be maximum illustrative and diagrammatic for comprehensibility on the concepts. The diagrams and the illustrations present should be self-explanatory, simple, clear and should satisfy the queries of the students. Illustrations in the form of cartoons and comic figures attract the students towards the illustrations and make them interesting (some examples are given in Appendix G). More of interesting illustrations should be incorporated in the content. Newspaper clippings can also be given in the text related to a particular concept, which will arouse curiosity among students and make them aware of the problems faced by the society as a whole (as given in

Appendix G). For example, in the lesson of water, newspaper cuttings regarding the shortage of water can be given. In the lesson pollution, newspaper cutting related to health concerns can be given. Illustrations given should be colourful and big in size as they attract more fascination and appreciation from the children. The illustration given in Class VI textbooks was black and white and small in size. Classes VII and VIII textbooks have bigger size illustrations but they were neither colourful nor labelled properly. Class VI textbook diagrams specially, lack nomenclature, for example, Fig. 1.1 of first lesson shows measuring instruments without any labelling. Figures 2.1, 2.5, 2.12, 3.1, 3.9, 5.5, 5.7, 5.15, 6.1, 7.7 and 11.2 were similar such diagrams in the textbook which were not labelled properly and require better explanation and elaboration in the text.

There were very few activities given in all the three textbooks. The few activities, which are mentioned, do not have clear portrayal of whether the activities are to be performed by the students or teachers. The books should be designed and prepared according to the philosophy of science and technology. The activities should be categorized into students' activities and teachers' demonstration. The list of minimum number of necessary items should also be given at the end of each unit so that teacher can acquire them prior to taking the class. There should be main points given at the end of each unit to recapitulate the contents of the unit. The main points could be represented either in bold or with different colour representation. The main points given in the main areas in the lesson, help the students in revision and help the teacher to design and prepare for the class.

The questions given at the back of the unit were satisfactory. Units have all kinds of questions that were short, long and objective type. Questions should also be given at the end of the concept as well as during the procedure of the activities.

In addition to this, in the beginning, before the introduction of the lesson the textbooks should have a flow chart showing the links between the units. The grid of processes and skills to be developed in the students and set of safety rules for the laboratory should also be given in the beginning of the textbooks (for example, as given in Appendix G). At the end of the textbook diagrammatic representation of various skills like using a burner, heating, boiling, filtering, measuring, drawing graphs, using a microscope etc. should be given diagrammatically (for example, as given in Appendix G). For each textbook the set of skills mentioned should be according to the content. Certain science words used in the text, with their meanings could be given at the end of the textbook.

Researchers felt that some supplementary reading materials might be made available to students to create interest in science. Some of the concepts, which were not understood or internalized by most of the students, were likely to become hindrance in further studies; it may develop a dislike or hatred for science learning. These concepts can be made simpler by providing supporting material. Supplementary material not only makes difficult concepts simpler but also makes science more interesting. Such material should not be bound by the limitations of the textbooks and should be of a different kind.

Teacher should also be provided with teacher's handbook, which would help in proper transaction of the content. Some of the activities and concepts require a lot of explanation which cannot be provided in the textbook as it would increase the content load, make the book bulkier and may be beyond the comprehension of the students. Such concepts and activities, which will help the teacher in the better transaction of the content, should be provided in the form of teacher's guide or handbook.

Analysis of the Content of the Textbook in terms of the Identified Themes of STS Framework in Science Education

Before embarking on the new STS curriculum it is important to examine the already existing textbooks in terms of the weightage assigned to various aspects of the themes of STS framework in science education. Reflection and analysis become instrumental to the process of infusing STS themes into existing text.

Rosenthal (1984) reported that over the period of twenty years from 1963 to 1983, the emphasis on science and society decreased in the twenty-two biology textbooks, used in the United States at BSCS level, that were examined. The percentage of the total text devoted to social issues in these textbooks ranged from 10 per cent to 47.9 per cent. Soong and Yager (1993) analyzed eleven textbooks, most frequently used at secondary stage in the United States, to determine the attention given to STS topics. Comparisons were made of STS coverage by science disciplines and across the seven to twelve grade levels. Results revealed that less than 7 per cent of the narrative space is devoted to STS topics with a range of 11.5 per cent to 0.5 per cent. Further, the coverage of STS topics decreased as grade level increased.

A great discrepancy was found to be existing between goals of science education (an STS emphasis) to actual coverage in the most frequently used textbooks. These studies clearly indicated the importance of analysis of the textbooks in terms of coverage of STS themes. These researches also support the need to analyze the existing text of the textbooks most widely used by Indian students at upper primary stage in terms of whether weightage is given to STS themes in the textbooks.

The textbook writers at the SCERT, Gurgaon and NCERT as well as science teachers of Haryana, during informal interaction, revealed that there was no set criteria for giving weightage in terms of percentage to various domains while writing the textbooks or for the time to be spent on the various domains during the transaction of the content. However, efforts have been made to give weightage to various assessment domains, for example, Northern Examinations and Assessment Board gave weightage in their syllabus to the domains for science, technology, society for GCE (General Certificate of Education) course (1996-97). The weightage given to the assessment objective in the syllabus is, 40 per cent to knowledge and understanding, 30 per cent to application and analysis and 30 per cent to communication.

The teachers and textbook writers of SCERT and NCERT, when informally asked as to what they felt should be the weightage given to various domains during the writing of the textbooks with an STS orientation, gave the percentages for the three classes VI, VII and VIII, as given in Table 6.1.

Table 6.1: Weightage of Domains in the Textbooks

S. No.	*Domain*	*Percentage Weightage in Different Classes*		
		VI	*VII*	*VIII*
1.	Acquisition of knowledge	35	30	27
2.	Skill and processes	20	22	25
3.	Ethics, Values and Attitudes	15	15	15
4.	Application	30	33	33

This weightage was used as a criteria for the interpretation of the results of the analysis of the textbooks of classes VI, VII and VIII.

The researcher analyzed the textbooks of the Haryana Education Board used at the upper primary stage (classes VI to VIII) in terms of coverage of STS theme in science education as validated for the present study. The analysis of the content was done sub-unit wise as already numbered in the textbooks. Each lesson in the textbook was divided into sub-units which in turn was marked by a specific number, for example, Lesson I of Class VI textbook has sub-units marked with numbers as 1.1, 1.2, 1.2.1 etc. Each sub-unit with the number was considered as a single entity for analysis. The analysis was carried out separately for each sub unit on the STS framework sheet and frequency counts were put for the various aspects in the different domains of the framework covered in each sub-unit. This exercise was done for each lesson of the textbooks for class VI, VII and VIII. A total of forty-eight lessons were analyzed, fourteen lessons of class VI, sixteen lessons of class VII and eighteen lessons of class VIII.

The lessons when analyzed in terms of the activities present in the textbooks showed that there were certain activities suggested in the syllabus which were incorporated in the text of the lessons. There were a total of ten activities in all mentioned in the fourteen lessons of class VI textbook. Not even an average of one activity per lesson was there in the textbook. Class VII textbook had a total of ten activities for sixteen lessons. Here again some lessons were without any activity. Class VIII textbook had a total of fifteen activities for eighteen lessons which was also too few. The activities given in all the three textbooks were not elaborated properly with the procedure and materials required for the activities. Some of them were in the form of showing chart and model only. The activities did not involve any technological skill or processes. Emphasis in the text should be more on the activities, which will help in inculcating values and a positive attitude towards science. The figures given in all the three textbooks were good in number but they were not very demarcating or clear. There was no diagrammatic representation of the activities given in the text. Some figures were showing the use of technology in daily life like electric meter, balance, bicycle, wind mill, crane, ball bearing, hand pump etc. but their parts had not been clearly labelled. More figures showing the use of science and technology in daily life should be explicitly given and figures on skill and processes, with proper procedure of holding various instruments, should be given.

The lessons were further analyzed on the frequency chart of the STS framework in science education. The frequency count for each aspect of the STS theme framework in science education was totalled sub-unit-wise and then lesson-wise counts of each aspect were tabulated. The detailed tabulation of each class is given in Appendix H. The data was tabulated separately for classes VI, VII and VIII. Percentage was calculated for the total frequency count of each aspect of each domain covered in all the lessons of each class. These percentages obtained for each aspect in a domain were then totaled domainwise. The tabulation done separately for classes VI, VII and VIII can be summarized further as given in Table 6.2. Table 6.2 can be analyzed both class-wise as well as domainwise. The graphical representation of the Table 6.2 class-wise is depicted by bar graphs number 6.1, 6.2 and 6.3 and the graphical representation of the table, domain wise, is depicted by bar graphs 6.4, 6.5, 6.6 and 6.7.

Table 6.2: Domain-wise Analysis of the Science Textbooks of Classes VI, VII and VIII

	Domain	*Percentage*		
		Class VI	*Class VII*	*Class VIII*
	Acquisition of Knowledge			
	Concepts related to Conventional Content of Science			
1.	Facts	5	1.3	1.5
2.	Information	14.9	17.4	15.9
3.	Concepts	8.6	5.1	5.7
4.	Laws	0	0.98	1.3
	Total	28.5	24.78	24.4
	Concepts related to Science and Technology			
5.	Interaction and Change	2.9	2.1	1.9
6.	Growth and Cycles	3.3	1.54	1.9
7.	Patterns and Processes	6.5	9.5	7.5
8.	Probability and Prediction	0.47	0.42	0
	Total	13.17	13.56	11.3
	Concepts related to STS			
9.	Describes Interaction of Science, Technology and Society	1.6	2.1	2.6
10.	Understands limits of Science and Technology	0.28	0	1.0
11.	Distinguishes Personal Opinion from Scientific Evidence	0	0	0.23
12.	Understands Application of Technology	1.1	1.5	3.0
	Total	2.98	3.6	6.83

(*Contd.*)

Table 6.2 (*Contd.*)

Domain	*Percentage*		
	Class VI	*Class VII*	*Class VIII*
Skill and Processes			
13. Questioning and Searching	6.1	7.3	2.8
14. Observing, using Space-Time Relationship	1.2	0.98	0.35
15. Classifying, Grouping and Organizing	5.7	7.3	6.0
16. Interpreting Data	2.2	2.5	1.9
17. Comparing and Conserving	8.3	4.1	4.4
18. Critically Thinking Skills	1.1	1.7	0.7
19. Use of Technology in Practical Situation	2.8	2.1	5.6
20. Producing Mental Images	2	0	0
21. Combining Objects and Ideas in New Ways	0.28	0	0
22. Predicting Possible Consequences	1	0.7	0
23. Suggesting Possible Causes of a Given Situation	1	0.7	0.23
Total	31.7	27.38	21.98
Ethics, Values and Attitudes			
24. Values Democracy and Democratic Process	0.28	0	0
25. Identifying Values Bearing on Any Given STS issue like local issues, public policies, global problems	0.76	1.7	1.3
26. Developing positive attitude towards science in general, science in school and science teacher	0	0	0.12
27. Developing positive attitude towards oneself	0	0	0
28. Expressing personal feeling in a constructive way	0	0	0
29. Developing sensitivity to and respect for the feelings of other person	0	0	0
30. Making decisions about social environment issues and exploring arguments on various sides of an issue	0.66	1.3	2.3
Total	1.7	3.0	3.72
Application			
31. Seeing instances of scientific concepts in everyday life experiences	12.8	10.4	11.6
32. Understanding Science and Technology principles in household devices	2.6	0.56	3.9
33. Using scientific process in solving problems in everyday life	4.2	4.9	5.62
34. Making decisions related to personal health nutrition and lifestyle based on knowledge of Scientific Concepts	1.7	4.6	5.62
35. Extended school experience beyond classroom	0	1.1	1.24
36. Handling of apparatus	0.85	6.2	3
37. Development of low cost material	0	0	0
Total	21.15	27.76	31.9

Class wise Analysis of Different Domains

Class VI

As can be seen from bar graph 6.1, in the content of the class VI textbook, maximum weightage was given to the domain acquisition of knowledge, i.e. 44.7 per cent and within it also maximum importance was given to the knowledge related to conventional content of science, i.e. 28.5 per cent and then to knowledge related to science and technology i.e. 13.2 per cent and lastly to knowledge related to science, technology and society i.e. 3 per cent. Skill and processes had a 31.7 per cent weightage, which is a positive depiction as science without skill and processes is meaning less. The weightage given to application domain was 22.2 per cent and to domain ethics, values and attitudes was 1.7 per cent. The weightage given to application domain was less in comparison to the knowledge and skill and processes domain, which should be more. Ethics, values and attitudes have very negligible depiction though they should be inculcated to the maximum from as early a stage as possible.

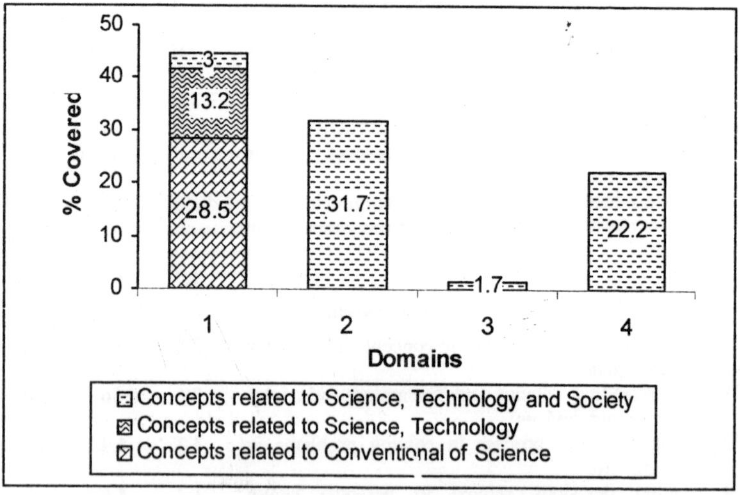

1. Acquisition of knowledge
2. Skill Areas and Processes
3. Ethics, Values and Attitudes
4. Application Area

Graph 6.1: STS Themes Covered under Different Domains in the Textbook of Class VI

Class VII

Bar graph 6.2 gives the representation of the content of class VII textbook under different domains. Here again the maximum representation was of the domain, acquisition of knowledge i.e. 42 per cent, then skill and processes i.e. 27.4 per cent, then application i.e. 27.8 per cent and least importance was given to ethics, values and attitudes i.e. 3 per cent. Within the domain, acquisition of knowledge, maximum coverage was of the content related to the conventional content of science i.e. 24.8 per cent, then to science and technology i.e. 13.6 per cent and least to the knowledge related to science, technology and society interaction i.e. 3.6 per cent.

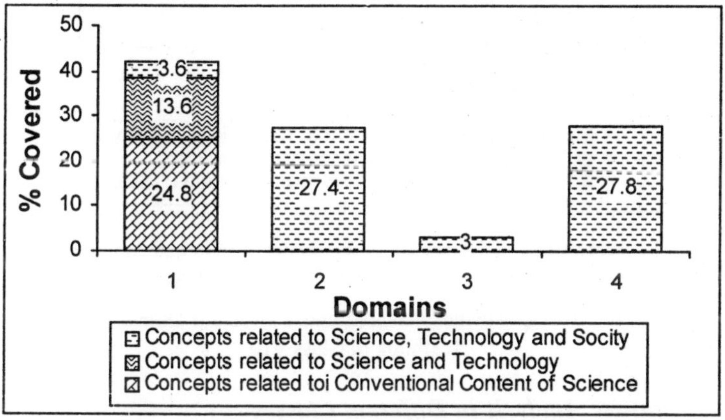

1. Acquisition of knowledge
2. Skill Areas and Processes
3. Ethics, Values and Attitudes
4. Application Area

Graph 6.2: STS Themes Covered under Different Domains in the Textbook of Class VII

Class VIII

Bar graph 6.3 depicts the percentage of content of class VIII covered under different domains. The maximum coverage was of the domain, acquisition of knowledge, i.e. 42.5 per cent. The domain application area comes next, i.e. 31.9 per cent which was different from the depiction in the classes VI and VII. Skill and processes had a depiction of 22 per cent and least presentation, i.e. 3.7 per cent was of ethics, values and attitudes. Within the domain, acquisition of knowledge, major thrust of the content was on knowledge related to conventional content of science, i.e. 24.4 per cent then to knowledge of concepts related to science and technology, i.e. 11.3 per cent and least to knowledge of concepts related to science, technology and society, i.e. 6.8 per cent.

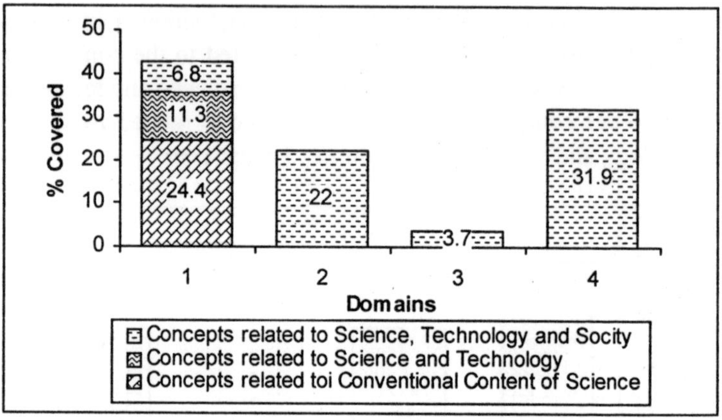

1. Acquisition of knowledge
2. Skill Areas and Processes
3. Ethics, Values and Attitudes
4. Application Area

Graph 6.3: STS Themes Covered under Different Domains in the Textbook of Class VIII

Domainwise Analysis in Different Classes

Acquisition of Knowledge

The Acquisition of knowledge related to science, technology and society is fundamental to science education. The domain further includes three sub-domains, i.e. knowledge of concepts related to conventional content of science, knowledge of concepts related to science and technology and knowledge of concepts related to science, technology and society. Analysis of Table 6.1 and its graphical representation, i.e. bar graph 6.4 depict that out of the total content, the content that emphasized on acquisition of knowledge domain of classes VI, VII and VIII is: 44.7 per cent: Class VI; 42 per cent: Class VII; 42.5 per cent: Class VIII. The results indicate an information-oriented text. A more detailed picture of science as a body of acquisition of knowledge is given in Table 6.2 where frequency of occurrence of each sub category has been recorded. Analysis of the sub-categories of this domain showed that the lessons emphasized more on conventional content of science, i.e. facts, information, concepts and laws (28.5%: Class VI; 24.8%: Class VII; 24.4%: Class VIII) and then on knowledge of concepts related to science and technology, i.e. interaction and change occurring in technological process, the various conceptual growth and cyclic changes in technology, patterns and processes of technology, and the probability and prediction of their sustainability (13.2%: Class VI; 13.6%: Class VII; 11.3%: Class VIII) and least importance was given to knowledge of concepts related to science, technology and society, i.e. interaction of science, technology and society, understand the limit of science and technology, distinguish personal evidence and understand application of technology (2.9%: Class VI; 3.5%: Class VII; 6.6%: Class VIII).

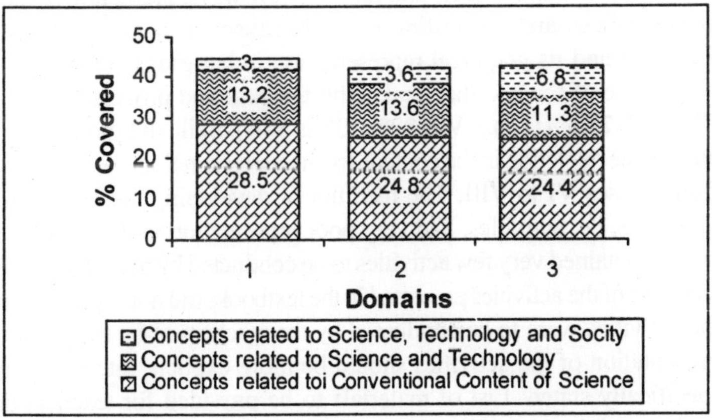

Graph 6.4: Classwise Comparative View of Coverage of the Domain, Acquisition of Knowledge

Though there was slight increase from Class VI to VIII in the last category, it was too less in comparison to the other two categories. Within this domain also the knowledge related to conventional content of science dominates over the knowledge related to science, technology and society, though a reverse of it should be depicted in the content, according to the philosophy of STS. The concepts related to science, technology and society had been completely neglected in the text of entire lessons at different levels. Aspects like distinguishing personal opinion from scientific evidence, understanding limits of science and technology and probability and prediction were either negligibly depicted or were totally missing from the textbooks.

The class VI content was more dominated by the acquisition of knowledge in comparison to that of class VII and VIII. This analysis depicts the need for making the content more STS oriented.

Skill Areas and Processes

The second category spells out science as a way of investigation and includes skills and processes aspects like questioning and searching, observing using space time relationship, classifying, grouping and organizing, interpreting data, comparing and conserving, critically thinking skills, using technology in practical situations, producing mental images, combining objects and ideas in new ways, predicting possible consequences and suggesting possible causes of a given situation. Table 6.1 and its graphical representation in bar graph 6.5 depict the percentage of the text that deals with the skill and processes (31.7%: Class VI; 27.4%: Class VII; 22%: Class VIII). Skill and processes had an average depiction in the content and its representation was decreasing from classes VI to VIII. The medium of developing these skills and processes, i.e. activities, had very poor presentation in the books. The books contained very few activities to be conducted by the students. The language of the activities presented in the textbooks did not depict whether the activities were to be conducted by the teachers or by the students. Presentation of the activities should be more explicit, interesting and specifically stated. List of materials to be provided for teachers and student activity should be clearly mentioned. The weightage given to the activities gets reduced from class VI to class VIII, which should not be so. The processes and skills like producing mental images, combining objects and ideas in new ways, predicting possible consequences and suggesting possible causes of a given situation are negligibly depicted or completely missing from the activities. The thrust of the activities should be on making science relevant to the immediate environment and needs of the students. Hence activities need to address the immediate problem of the environment like population, health, use of weedicides, pesticides and fertilizers, use of energy, and use of technological knowledge in their daily lives. Skill area and processes should have more importance and equal depiction through the conduction of activities in all classes and the variance should be from simple to complex in terms of the skills to be developed from classes VI to VIII, which was not so in the text.

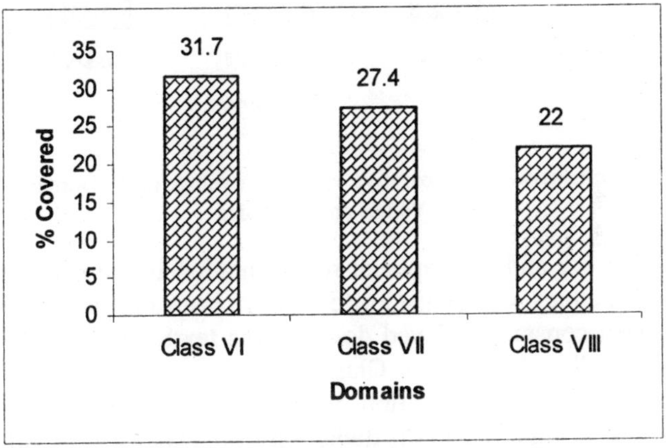

Graph 6.5: Classwise Comparative View of Coverage of the Domain, Skill and Processes

Ethics, Values and Attitudes

The third category dealt with a very important aspect, which was the crux of the whole philosophy of science, technology and society and a growing need of today's society i.e. development of positive ethics, values and attitudes. It includes valuing democracy and democratic process, identifying values bearing on any given STS issue like local issues, public policies and global problems, developing positive attitude towards science in general, science in school and science teacher, developing positive attitude towards oneself, expressing personal feelings in a constructive way, developing sensitivity to and respect for the feelings of other people developing attitude, towards and appreciation for technology and making decisions about social environment issues and exploring arguments of various sides of an issue.

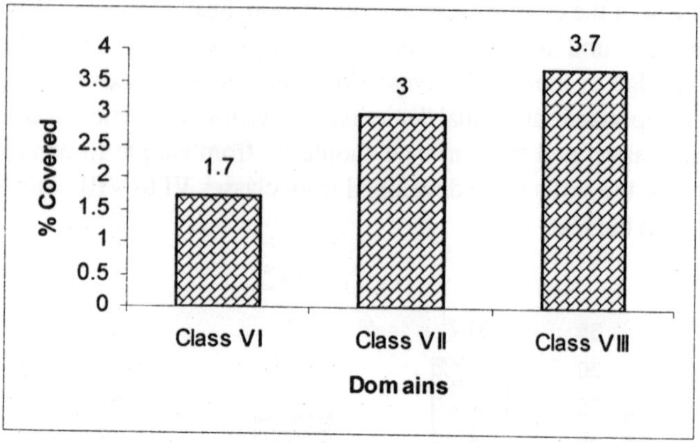

Graph 6.6: Classwise Comparative View of Coverage of the Domain, Ethics, Values and Attitudes

Table 6.2 and its graphical representation in the bar graph 6.6 show that ethics, values and attitudes have been a neglected area and have been constantly played down in the textbooks (1.64%: Class VI; 3.6%: Class VII, 2.84%: Class VIII). Ethics, values and attitudes have a very poor presentation, which was even below 4 per cent. There was a slight rise in the depiction from classes VI to VII, but it decreased in class VIII, though values should be inculcated from as early a stage as possible. Some of the aspects under this domain, like valuing democracy and democratic process, developing positive attitude towards science in general, science in school and science teacher, developing positive attitude towards oneself, expressing personal feelings in a constructive way, developing sensitivity to and respect for the feelings of other people, were either totally missing or had been negligibly represented in the text. Attention needs to be focussed on the effects of science and technology upon values, ethics and attitudes. Textbooks should assist the individual to develop a coherent system of values in order to live and survive in the modern world, should help in valuing the positive aspect of technology and such issues should be addressed emphatically in the science curriculum at the upper primary level.

Application Area

The percentage of text stressing on application area was slightly higher than the previous domain as can be seen from the Table 6.2 and its graphical representation depicted by bar graph 6.7 (21.4%: Class VI; 26.7%: Class VII; 31%: Class VIII). There was a rise in its depiction from classes VI.to VIII, which was a positive aspect but further analysis revealed that development of low cost materials was totally neglected in all the three textbooks. Further extending school experience beyond class rooms, handling of apparatus, understanding science and technology principles involved in household devices and making decision related to personal health, nutrition and lifestyle based on knowledge of scientific concepts, have been poorly depicted in the content. There were very few instances giving the positive and negative effects of science and technology or a discussion on the related societal issues. The content covered under this area should be more local specific as one of the major themes of science, technology and society education is that the child should be able to apply the scientific and technological knowledge locally in everyday life experiences. More importance should be given to the application of technology in daily life, portraying the positive as well as the negative effects of technology, giving importance to upcoming problems and challenges in the rural areas related to health, nutrition, population, agriculture and education.

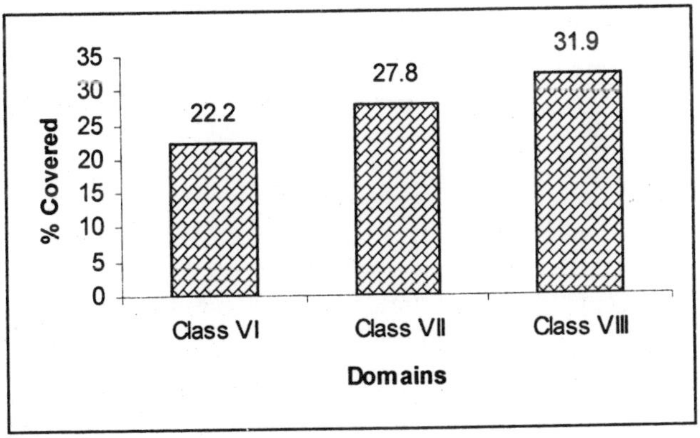

Graph 6.7: Classwise Comparative View of Coverage of the Domain Application Area

These representations clearly show that in all classes, maximum importance was given to acquisition of knowledge and the least importance was given to ethics, values and attitudes. Skill and processes and application were averagely covered in the text. The STS approach lays more emphasis on the skill areas and processes, application area and ethics, values and attitudes. Ethics, values and attitudes have been totally neglected and the content needs to be revamped in this context. The whole analysis can be summarized as: the text was more information oriented, there was little emphasis on science as a way of thinking and the relevance of science to daily life. There were very few instances giving the positive effect of science and technology or a discussion of the related societal issues. Negative effects of technology had also been not referred to. The students aged 11 to 14 years of the upper primary level are at the transitional phase between the concrete operational and formal operational stages as described by Piaget. In this intermediate phase the students require more emphasis on skill and processes, and values and attitudes than had been provided in the text. The right balance can occur only if stress is laid on reducing the content and increasing the number of activities for inculcation of values and skills.

7

Incorporation of STS Input in Textbooks and its Transaction: Possibility and Impact

Introduction

STS framework in science education had been designed in an effort to interpret science and technology as complex socially embedded enterprises, and to promote the development of a critical, scientifically and technologically literate citizenry capable of understanding STS issues, empowered to make informed and responsible decisions and able to act upon those decisions. What remains problematic is the translation of this framework into the existing instructional material and instructional practices. There are two possibilities of addressing this problem. One option is to infuse STS material, modules or units as in SATIS (Science & Technology in Society Project Launched in Britain in 1986) and into existing course as in S-STS (Science through Science, Technology and Society developed in the USA in 1985) into the existing course. The other is to develop separate, freestanding STS courses (such as Science in Society, developed in Canada in 1988; or SISCON, Science in Social context, initiated in Britain in 1983).

The modification of the existing courses to accommodate STS perspectives is the option that the researcher chose. This view of integrating the STS theme is supported by much of the related literature (Aikenhead and Ryan, 1992; Heath, 1992; Bybee, 1991; Solomon, 1993; Ramsey, 1994; Thier & Nagle, 1994) and by the proliferation of curriculum development projects. (SATIS and S-STS). Piel (1990) in his recommendations for science/technology/ society curriculum suggests that existing courses of study should be evaluated and material, which is obsolete or not relevant, should be deleted. This would provide 'space' to infuse into existing science programmes selected STS topics. Infusion of STS related topics into courses that are required at a particular level would ensure that the total spectrum of the student population would be exposed to the important area.

Analysis of Textbooks in Terms of Possibility of STS Incorporation

The content of the textbooks was analyzed keeping in view the previous analysis of the content in terms of the themes of the STS framework in science education. The analysis of the content was done sub-unitwise as already numbered in the textbooks. Each lesson in the text book was divided into sub-units which in turn was marked by a specific number. Each sub-unit is the number that was considered as a single entity for analysis. The analysis was carried out sub-unit wise under three headings, namely number of sub-units with the possibility of incorporation of STS, number of sub-units having STS theme already present and number of sub-units not requiring STS input as the content fulfils other objectives of science education, particularly required for higher learning. The analysis was carried out sub-unitwise for each class and tabulated as given in Tables 7.1, 7.2 and 7.3.

Rather than reconstructing the entire science curriculum around "social issues" the researcher first analyzed the existing content in terms of the possibility of incorporation of the STS theme and then modified two lessons of the science textbooks for class VI, VII, & VIII. This was done by integrating STS themes into the existing content, wherever there was maximum possibility of incorporation. The researcher tried to spell out the additional input in terms of: (i) activities, (ii) diagrammatic representation, (iii) application from daily life, (iv) skills and processes, (v) value and attitude, (vi) conceptual clarity, and (vii) field visits. Integration of the STS themes was not done by adding new concepts to the text but by widening the already existing concepts in lessons by exemplifying them in terms of social relevance and interest of students. An emphasis upon increasing process skills that students can use in resolving their own problems, creatively developing activities using different teaching strategies like simulation, games and role-play and upon use of local resources to locate information that can be used in problem resolution.

Table 7.1 shows the analysis of the class VI textbook. It clearly shows that there was a possibility of STS incorporation in all the lessons though it varies from 11 per cent to 73 per cent. The lessons Measurement and Water have the maximum possibility of incorporation of STS in the content. The researcher modified these two lessons. The rest of the lessons also have the possibility of incorporation of STS themes in the content but it was not possible

Table 7.1: Overall Sub-unitwise Analysis of Lessons of Class VI Textbook with the Possibility of Incorporation of STS—Whether already Present or Whether not Required

Lessons	*Title*	*Number of Sub-Units*				
		Incorporation of STS Possible	*STS Already Present*	*STS not Required*	*Total Number of Sub-units*	*Incorporation of STS Possible (in terms of %)*
1.	Measurement	11	3	2	16	68
2.	Things around us	5	2	5	12	42
3.	Separation of substances	9	1	3	14	64
4.	Changes around us	4	4	1	9	44
5.	Speed, Force and Pressure	5	9	3	17	29
6.	Simple Machines	2	6	3	11	18
7.	Universe	5	1	17	23	21
8.	Living World	1	1	7	9	11
9.	Plants and Human, Composition and Work	5	2	9	16	31
10.	Food and Health	4	2	5	11	36
11.	Balance in Nature	4	5	4	13	31
12.	Balance in Environment	4	1	1	6	67
13.	Water	8	1	2	11	73
14.	Energy	10	5	3	18	56

for the researcher to modify all. There are some lessons like Speed, Force and Pressure and Simple Machines which already have some element of STS coverage in the content but these elements are only restricted to quoting examples. The content of the lessons could have been made much more useful if some child-centered activities were also included. For example, where the content of these lessons gave examples of a screwdriver, scissors and hammer, there some activities could have been included giving the uses of these simple machines/instruments in their daily life. Students would also simultaneously acquire the process skills for their proper usage. While giving the example of boxes working on hinges, mention could have been made of how screwdrivers can help in repairing hinges. Where a balance was given as an example, its proper use and how students can observe its usage by shopkeepers could have been mentioned so that they do not get cheated. Though the lessons having sub units with STS themes like Separation of Substances, Balance in Nature and Energy cover the application domain of the STS framework, the skill and ethics, values and attitude domains are missing.

Table 7.2 gives the analysis of the lessons of the class VII textbook. Here again, the units were analyzed under the already discussed categories. The percentage of possibility of STS incorporation in the sub units varied from 13 per cent to 75 per cent. The researcher has modified lessons with maximum possibility of incorporation, i.e. Population and Pollution. These two lessons were also taught to the experimental and control group during the experiment.

The lessons, which already had STS, coverage like Transmission of Heat, Diseases and their Control and Animal-Care, also had scope for further modification by creatively formulating activities which would enhance skill areas, processes, ethics, values and positive attitudes. For example, in the lesson of Diseases and their Control, diseases with the latest threat to the society like HIV, T.B. and Polio should be discussed. Helping each other and society at large to fight the diseases, valuing a clean environment and various precautions to prevent diseases could be discussed in the content. Practical exercises on preparing an emergency kit or first aid box for their home, class and school could have been done. Certain other lessons like Pressure and Buoyancy, Nature and Composition of Substances and Your

Own Protection had a major thrust for knowledge. Some of the concepts in these lessons like latent heat of fusion, vaporization, laws of conservation of mass, valiancy etc. are difficult in terms of comprehension for standard VII these concepts could either be dealt in higher classes or alternatively could have been given in a simpler form.

Table 7.3 gives the sub-unitwise analysis of the content of the class VIII textbook. On analysis it was found that the percentage of possibility for incorporation of themes of STS ranged from 5 per cent to 50 per cent. This percentage is less than the percentages of classes VI and VII. This is probably due to two reasons; firstly, some of the lessons have content which do not require STS input as they with other objectives of science education. The book was loaded with lessons like Potential Difference and Resistance, Electromagnet, Composition of Atom, Cell and Tissue and Liquid whereas, the content deals with concepts of higher learning. These lessons were theoretically sound. However, the presentation could have been made more interesting by citing some examples, with comic representation or with other teaching strategies. Secondly, there were certain lessons like Electrical Energy, Carbon and its Uses, Acid, Base and Salts, Crop Cultivation, Useful Plants and Animals and Science in the Use of Man, which have a fair incorporation of STS in their content, but content lacks practical exercises. For example, in the lessons of electrical energy, the electric meter is described but the skill needed by students to read it is not taught. The value of saving electricity could be inculcated through activities starting from their school premises. They could be made aware of appliances that consume a lot of energy, which would further strengthen the value of saving energy. Similarly, in the sub-units which describe the transmission of electricity with the help of transformers, stealing of electricity and its bad effects could be discussed. Value of not stealing the electricity and paying the bills on time could be discussed.

Similar incorporations were also possible in the other lessons that have content with STS themes. Practical aspects were only discussed theoretically. There were no value-based activities, which should also be incorporated. The other lessons like Light Refraction, Vibration, Speed and Sound, Our Living World, Growth and Development, Agriculture and Conservation of Natural Resources had a strong possibility of incorporation of STS in the content. Lessons with maximum possibility of incorporation of themes of STS like Light Refraction and Agriculture have been modified by the researcher.

Table 7.2: Overall Sub-unitwise Analysis of Lessons of Class VII Textbook with the Possibility of Incorporation of STS—Whether already Present or Whether not Required

Lessons	*Title*	*Number of Sub Units*				
		Incorporation of STS Possible	*STS Already Present*	*STS not Required*	*Total Number of Sub-Units*	*Incorporation of STS Possible (in terms of %)*
1.	Motion and Mass	4	1	5	10	40
2.	Pressure and Buyouancy	2	2	9	13	15
3.	Heat and its effects	6	4	1	11	55
4.	Transmission of heat	1	4	0	5	20
5.	Light	3	1	7	11	27
6.	Static Electricity	2	1	3	6	33
7.	Effect of Electricity	2	0	1	3	67
8.	Magnetism	2	1	2	5	40
9.	Nature and constitution of matter	7	1	13	21	33
10.	Air and oxygen	2	1	4	7	29
11.	Water	5	1	7	13	39
12.	Protection of the self	4	0	6	10	40
13.	Diseases and their control	1	6	0	7	14
14.	Population	3	0	1	4	75
15.	Pollution	5	0	2	7	71
16.	Animal care	1	7	0	8	13

Table 7.3: Overall Sub-unitwise Analysis of Lessons of Class VIII Textbook with the Possibility of Incorporation of STS. Whether already Present or Whether not Required

Lessons	*Title*	*Number of Subunits*				
		Incorporation of STS Possible	*STS Already Present*	*STS not Required*	*Total Number of Sub-units*	*Incorporation of STS Possible (in terms of %)*
1.	Light Refraction	9	4	10	23	39
2.	Potential Diff. and resistance	2	0	5	7	29
3.	Electrical Energy	1	3	3	7	14
4.	Electro-magnetism	1	2	4	7	20
5.	Constitution of an atom	1	1	6	8	13
6.	Vibration, speed and sound	3	0	4	7	43
7.	Carbon and its uses	3	5	8	16	19
8.	Acid base and salt	2	3	4	9	22
9.	Our living world	4	1	4	9	45
10.	Cells and tissues	4	0	5	9	45
11.	Reproduction	1	1	2	4	25
12.	Growth and Dev.	2	1	3	6	33
13.	Material	3	1	6	10	30
14.	Agriculture	6	5	2	13	46
15.	Growing of crops	2	4	6	12	17
16.	Useful plants and animals	4	8	4	16	25
17.	Conservation of natural resources	3	2	2	7	43
18.	Science in use for man	1	1	0	2	5

Modification of Two Lessons Each of Classes VI, VII and VIII Science Textbooks in Terms of the Possibility of STS Incorporation in the Existing Content

Two lessons each of classes VI, VII and VIII were modified in terms of the possibility of STS incorporation in the already existing content. The lessons chosen for this purpose were based on the previous analysis already discussed. The lessons analyzed to be having the maximum scope of STS incorporation were modified.

There were other lessons also with the possibility of incorporation but it was not feasible in the present study for the researcher to modify all the units. Within the lessons also only those sub-units were modified where there was possibility of STS incorporation. The other sub units were not discussed. Lessons which were modified are given below in Table 7.4. The incorporations made in these lessons are presented in Tables 7.5, 7.6, 7.7, 7.8, 7.9 and 7.10.

Table 7.4: Name of the Modified Lessons

S. No	*Class*	*Name of the Lesson*
1.	VI	Measurement
2.	VI	Water
3.	VII	Population
4.	VII	Pollution
5.	VIII	Agriculture
6.	VIII	Light refraction

Table 7.5: Measurement (Class VI)

Sub-unit No.	*Concept*	*Suggested Content*	
		Area	*Actual Incorporations*
1.1	Measurement we make in everyday life	Diagrammatic representation	• Of measurement. with respect to the society, i.e. *tarazu* used by vendors, measuring can used by milkman, in petrol pumps, in kerosene depots or Instruments used by scientists. • Relating to the comparison of heights and classrooms.
1.2.1	Some simple methods of measurement	Activity	• Estimating the length of a table and checking by accurate measurements. • In the absence of standard weights, weights written on soaps, paste etc. used as measuring objects or for comparison.
		Examples from daily life	• Of local methods to measure and how these vary from person to person like arm length, foot length etc. • Relating local units to standard units like *dhadi* (धड़ी), *man* (मन), *dhela* (ढेला) kilo (किलो) etc.
		Conceptual clarity	• Defining and explaining the rational of standard units.

(Contd.)

Sub-unit No.	*Concept*	*Suggested Content*	
		Area	*Actual Incorporations*
	Common scale and Standard scale	Examples from daily life	• Where we use the common scale in daily life and where it is necessary.
1.3.1	Length [meter]	Activities	• On making your own scale with graph paper • On three different scales varying in their units with diagram and with the relation of a higher unit with lower unit (mm, cm, m, km). • On measuring the height of ten pupils and representing them on a graph paper in a descending order studying the graph to find the average height.
		Skill Processes	• Of using a scale accurately along with the help of a diagram (the correct position of the eye while measuring).
		Values Attitudes	• Being cautious before buying anything (checking whether the scale used by the dealer is correct, stamped by the department of weights and whether the measure used by the shopkeeper is with arrow marks (→←) on both ends.)
		Application in Daily Life	• Reading Milo meters of cars, buses and trucks.
1.3.2	Mass (kg)	Examples from Daily Life	• Mass expressed in grams, kilograms and pounds with examples of the weight of a new born child in pounds, grown up boy in kilograms and weight of substances like sugar, salt, paste, wheat, bajra etc. in both grams and kilograms. • Some multiples of kilograms with examples.
		Diagrammatic Representation	• Of measurement of mass with spring balance and beam balance with diagram.
		Field visit	• To the grain depot to see the buying and selling of large quantities of grains, which the farmers grow (with useful tips that can be used by the farmers so that they do not get cheated).
		Activity	• On measuring of some amount of sugar or common salt or any other substance by means of beam balance using various weights ranging from one gram to one kilogram.

(Contd.)

Sub-unit No.	Concept	Suggested Content	
		Area	Actual Incorporations
		Attitudes and Values	• Consumer education regarding weights while buying various commodities.
1.3.3	Time (seconds)	Diagrammatic Representation	• Of various types of clocks.
		Skill and Processes	• Of handling the clock, concept of stopwatches and their working (generally in sports activities) to be discussed. • Of reading of railway timetable in hours and minutes and skill of reading the digital watch at a railway station.
		Conceptual Clarity	• Concept of time taken to form a minute, an hour, a day, and a year.
		Value Attitude	• Various activities in daily life which demand punctuality.
1.3.4	Degree Celsius	Skill Processes	• Process to measure the body temperature.
		Diagrammatic Representation	• Of various types of thermometer, measuring the temperature of the body, air and water.
		Activities	• On measuring these using different thermometer and the skill of reading the thermometer correctly.
		Application in daily life	• Of reading the temperature in a newspaper.
1.3.6	Volume	Diagrammatic representation	• Of the method of measuring the volume of a solid and a liquid. • Of different volume measuring instruments like measuring glass, measuring cylinder, conical flasks, pipette and instruments used by oil dealers.
		Skill Processes	• Use of measuring cylinder, its proper handling and reading of the level of liquid with proper placement of eye on the gradations.
		Activity	• Of measuring average volume of a drop of water using a dropper and measuring cylinder.
		Application in daily life	• Of proper reading and usage of droppers given with the medicines. • Consumer education regarding measuring of volumes of different liquids used in daily life like petrol, diesel, kerosene, mustard oil and various cooking oils.

(Contd.)

Sub-unit No.	*Concept*	*Suggested Content*	
		Area	*Actual Incorporations*
1.4	Estimation and measurement	Activity	• To measure sugar with a teaspoon and on a beam balance. Estimating the weight of the given amount of sugar using the above teaspoon. Also estimating the weight of salt using the same spoon and comparing by actual weighing.
		Application in daily life	• Examples of using estimation and actual measurements in their day-to-day lives—e.g. Actual measurements are used for weighing gold, silver and precious stones, but for measuring food commodities in cooking we use only estimations, usually.
1.7	How to keep standard devices	Values and attitudes	• Necessary tips to be kept in mind for the authenticity of the standard devices while buying them. • Knowledge of where to complain and knowledge of procedure, if wrong measuring devices are being sold or used.

The lesson on measurement had eleven sub-units with the possibility of STS incorporation. It lacked diagrammatic representations, activities, application in daily life, skill and processes, value and attitudes and conceptual clarity of few concepts. Need was felt to incorporate examples from daily life like local measuring instruments and units and various activities related to the various concepts of measurement like measuring the weight of salt, sugar, measuring length of table, classroom etc. Diagrammatic representations were there in some of the sub units. However, diagrammatic representations related to various measuring devices were missing. Consumer education related values and attitudes were totally missing from the sub-units which could have been elaborated upon like being cautious while buying things, authenticity of the standard devices, punctuality etc. Skills and processes also need to be included for proper usage of measuring devices like scale, reading railway time table, process of measuring body temperature, using of measuring cylinder etc. Field visits to a grain depot, shop, railway station could be included.

Table 7.6: Water (Class VI)

Sub-unit No.	*Concept*	*Suggested Content*	
		Area	*Actual Incorporations*
13.2	Water is an essential entity for life	Value and Attitude	• Story of two farmers—one who stores water for irrigation and the other who does not and so to suffer.
		Application in Daily Life	• Listing of all the uses of water both nationally and in their daily life and the life of other organisms. • Listing of natural sources of water and categorizing them into fresh and saline. • Identifying and describing the two main water sources in Haryana.
		Diagrammatic representation	• Of various uses of water for man
		Conceptual Clarity	• Distinguishing surface water from ground water
		Value and Attitude	• Role of habit of saving water by closing the taps properly at home, school and public places. Giving complaints for leaking taps and faucets.
13.3	Water cycle	Diagrammatic representation	• Of water cycle.
		Activity	• Demonstration of the effect of temperature on the states of water and explaining how ground water is replaced.
		Skill and Processes	• Preparing flow chart of water cycle
		Value and Attitude	• Discussion on human interferences in the water cycle like digging of wells, handpumps, tanks for irrigation and pollution.
13.4.2	Density	Application in Daily Life	• Water used in many machines, cars and other vehicles. • Aquatic animals and plants benefit from this property.
13.5	Making water safe for use.	Diagrammatic representation	• Of how pollution of water takes place.
		Conceptual Clarity	• Discussion on how various natural and human acts pollute water.
		Assignment	• Projects to be done by students on water pollution.

(Contd.)

Sub-unit No.	*Concept*	*Suggested Content*	
		Area	*Actual Incorporations*
13.5.1	Ways of making hard water soft	Activities	• Collection of water from various sources like river, well, tap, pond etc. and testing them for their difference. • Testing the hardness of water from different sources and the amount of lather formed from different samples of water.
		Conceptual Clarity	• Using cartoon strips for demonstrating hard and soft water. • Discussions on why certain samples of water taste salty. • Discussion on unsuitability of hard water for washing, for industries especially where boilers are used because it makes the boiler's surface hard.
		Activity with Diagrammatic Representation	• On putting different types of salt like calcium chloride, sodium chloride, calcium sulphate, magnesium sulphate and sodium carbonate and observing the amount of lather formed when each salt is put separately in water (tap water). Identification of salts which make water hard. • Removal of hardness of water.
13.5.2	Making water safe for drinking.	Diagrammatic representation	• Of both the correct and wrong ways of construction of a well–Proper distance to be maintained between a well and other sanitation constructions or growing of plants, covering of wells and proper method of construction of walls around wells. • How ponds can be made safe for drinking by proper construction and by imposing restrictions. (Diagrammatic representation of the proper construction of a pond). • Proper construction of taps and bore wells with diagrammatic representation of the right and wrong way of construction. • Proper sewage system for outlet of water, taps connected to pipe water supply should have filtration unit with no faulty connection.
		Value and Attitude	• Small tips of keeping water safe for drinking at home, school and public places.

(Contd.)

Sub-unit No.	*Concept*	*Suggested Content*	
		Area	*Actual Incorporations*
		Activities	• Of different stages in the purification of town water supply. • On making your own solar still for water purification.
		Skill and Processes	• Common methods of purifying water at home. Training students with simple technologies for water purification with the help of community resources.
		Conceptual Clarity	• Discussion on personal, family and community hygiene; environmental methods to prevent pollution of water resources.
13.6	Water-borne diseases and their cure	Value and Attitude	• Story of a village where villagers believe that these diseases are caused by the anger of gods.
		Conceptual Clarity	• Of how intestinal infectious diseases are contracted, list the symptoms, and suggest how to prevent them.
		Skill and Process	• Of use of an oral dehydration drink and the method of making it.
		Assignment	• Getting information from the clinic about these diseases. • Discussing first aid to handle these diseases. • Proper sanitation procedures so that animal and human waste do not mix in water, proper cleaning of body parts and use of latrines.
13.7	Water as one of the main natural resources and its conser-vation	Diagrammatic Representation	• Person using water irresponsibly and people queuing with buckets at a stand pipe or bore well.
		Application in Daily Life	• Of appropriate technology for rainwater harvesting and water shed development. • Recent local newspaper clippings about the problem faced in the various regions of their state and other states.

(Contd.)

Sub-unit No.	*Concept*	*Suggested Content*	
		Area	*Actual Incorporations*
			• Building of huge tanks at different levels for storage of water and role of panchayat for its proper distribution. • Estimating the amount of water used for cattle and for farming. • Inferring from data the amount of water used for certain activities and water consumed per family.
		Value and Attitude	• Story of a farmer who suffered the problem of a dried up well. Concept of bore wells in the village from which everyone is benefited and peasants association to be set up for water to be distributed equally among the villagers.
		Skill and Processes	• Reading of water meter installed at home or in the farms and computation of amount of water used in a month and the cost of water consumed. • Methods of collecting and storing rain water. • Graphic representation of the increasing demand for water. • Repairing of any leaking faucet or tap.
		Field Visit Activity	• Visit to a dam or the community tank. • Planting of trees near tanks and dams. • On measuring the water consumed at your home and by your family members everyday.
		Conceptual Clarity	• Distinguishing the types of faucets used. • Other ways of conservation of water like purification and recycling of sewage water.

There were eight sub-units in this lesson with the possibility of STS incorporation. The maximum input was given in the form of diagrammatic representation, conceptual clarity, value and attitudes, activities, examples/application in daily life, skill and process and lastly assignment/field visits. Diagrammatic representations like various sources of water, appropriate and inappropriate way of construction of wells and water resources, the procedure for making hard water soft etc. would make the concepts more explicit. Conceptual clarity was required for distinguishing surface water from ground water, pollution of water from various sources, difference between hard and soft water. Values and attitudes regarding conservation of water resources, inculcating habits for saving water and keeping water safe for drinking need to be inculcated. Activities were totally lacking in the lesson. Activities which will help them make the water safe for drinking, making hard water soft, planting trees around dams and tanks and for conservation of resources need to be included. Skill and processes like preparing flow charts, purifying water, preparing oral dehydration solution, repairing leaking taps and faucets, reading water meters and skill of collecting water are some aspects which need to be included. Field visits to dams and various water resources could also be planned.

Table 7.7: Population (Class VII)

Sub-unit No.	*Concept*	*Suggested Content*	
		Area	*Actual Incorporations*
14.1	Introduction	Values and Attitudes	• Playing a game or role-play for introducing the concept of population. • Readings, films, televisions and discussions to show the effect of overpopulation on society, i.e. a small family with one child or two children will be economically better because their consumption per member will be higher which will lead to better standard of living. • A story of how a family of ten people sleeping in a closed room with a burning coal *angithi* were found unconscious in the morning; story of a farmer and as to how his acres of land went on getting divided with each generation • Through a set of questions observing student's view or awareness about the population. • Study the Indian situation in which population increases by every 1.5 second. India having the second largest population in the world falling behind China and contrasting it with the Chinese decision on the one child per family programme and its consequences.

(Contd.)

Sub-unit No.	*Concept*	*Suggested Content*	
		Area	*Actual Incorporations*
		Diagrammatic Representation	• Of the overgrowing population.
		Conceptual clarity	• Small family norm and large family norm explaining through role-play or simulation. • Having discussions with grandparents about how the place they live in was during their childhood and now.
14.2	Counting the population	Diagrammatic representation	• Of the pictographs of the different states. • Of pictographs or bar graphs to illustrate the population of their village district and state.
		Skill and process	• Of interpreting data from the pictographs.
		Activity	• Experiments of growing bacteria in closed environment demonstrating what happens when the population grows too fast. • Knowing the population of nearby villages and counting the population of their own village. • Game of clusters of stones to facilitate dealing with the relationship of birth and death rates to growth rate.
		Assignment	• Their observations of population growth, the number of houses in the neighbourhood, the approach roads, new streets or colonies, wild growth which has now been replaced by human habitation and any other such notable changes.
14.3	Factors affecting population	Value and attitude	• Films and simulation showing how the impact of technology advances on family unit. How latest machines have decreased the death rates and how lifting a ban on abortions have helped the control of population • Valuing the concept of a small family to overcome poverty, fatal diseases and sanitation conditions, pollution of natural resources, scarcity of jobs etc.

(Contd.)

Sub-unit No.	*Concept*	*Suggested Content*	
		Area	*Actual Incorporations*
			• Discussions on population control strategies that generate awareness of treating the birth of a girl or a boy equally, stating the ill-effects on the mother's health if she gives birth to many children, explaining the right age of the girl and the boy to get married. • Discussing researches which show that educated women have fewer children than uneducated women (a research shows that in Brazil uneducated women give birth to 6.5 children those with secondary education bear only 2.5 children). • Removing myths and beliefs that only the birth of a male child will bring about emotional, social, religious and economical fulfillment. • Removing wrong beliefs that the birth of a child is a gift of god.
		Conceptual Clarity	• Discussing the importance of population planning at the national level. • Impact of over population on the essential services (energy, transport and health care supplies). • Showing films on how improved transportation has resulted in population shifts in the sub-urban regions of Haryana. • Discussing the effects of over population on the environment (common diseases in slum areas, water consumption of safe drinking has increased so much that water supply falls short of people's need etc.). • Natural resources like rivers and forests are getting polluted.
		Activity	• On how social evils can be controlled by reducing the population. • Discussions and demonstrations on rate of filtration of sewage to show the effects of high population density on the environment.
		Example from daily life	• Giving them live examples of how daughters are an asset to the family in terms of helping old age, social and economic security. Daughters like sons are capable of supporting parents in old age.

This lesson had three sub-units with the possibility of STS incorporation. There was need to incorporate positive values and attitudes about small families, removal of various myths regarding the birth of a girl child, technological advances on family unit, population control strategies and developing a positive attitude towards the birth of girl child. There was not even a single diagram in the content of this lesson. Diagrammatic representations of the overgrowing population and pictographs could have been included. Conceptual clarity regarding over population affecting environmental quality, importance of population at the national level is required. Activities and assignments regarding a population in the neighborhood and the lack of infrastructure due to increasing population need to be incorporated. Skill and processes regarding interpretation of pictographs and population data also needs to be incorporated.

Table 7.8: Pollution

Sub-unit No.	*Concept*	*Suggested Content*	
		Area	*Actual Incorporations*
15.2	Changes in the environment	Values and attitudes	• Story telling by the teacher along with pictures to explain how over population can affect the life of animals and plants. • Discussion about a clean or polluted environment and ways how a clean environment can be maintained and how a polluted one can be improved upon, e.g. students can do their duty by taking the rubbish out of the canal, sweeping the street and picking up bus tickets and pieces of paper.
		Examples from daily life	• Students give examples of an environment, they see and indicate whether it is clean or polluted. Discuss various examples from their experiences like polluted water in the canal, public parks with beautiful trees, streets littered with bus tickets and pieces of paper, neighbours turning on the radio aloud, digging a pond as a reservoir to store water for village use etc.
		Diagrammatic Representation	• A polluted environment in the community.

(Contd.)

Sub-unit No.	*Concept*	*Suggested Content*	
		Area	*Actual Incorporations*
		Activity	• Celebrating environment protection day in the school and the community. • Demonstration of dust pollution on plants (picking leaves from the same height from two plants—one on the roadside and from one which is in the interior of the village. Make a print of the leaves with gummed paper).
15.3	World full of noises	Assignment and field visit	• Visit your local hospital and meet the ear, nose, throat (ENT) specialist. Find out about the incidence of hearing loss amongst the population in your community. In particular ask about hearing loss among those who play in bands at weddings and other festive occasions and amongst traffic policemen. • Visit local police station and ask if they have registered case of nuisance under CRPC Section 133 against those causing noise pollution. If they have, find out details of the case and how it was ultimately disposed of, no such case has ever been filed and you feel that their has been noise pollution in your community ask the SHO at the police station why he has not filed such a case.
		Conceptual clarity	• Discussion of noise sources around them and the effect of high intensity noise on human being.
		Skill and Processes	• Noise mapping of your locality at different times during a day.
		Activity	• Designing and developing your own noise level meter. • Action plan for sound pollution control like making door to door survey for the awareness of noise pollution law, appealing to music system manufacturers to include the information of hazards of loud sounds in the text of their advertisements and planting of more trees and plants which have more stomata as the plants and trees with more stomatas absorb sound.

(Contd.)

Sub-unit No.	*Concept*	*Suggested Content*	
		Area	*Actual Incorporations*
			• Make your area green as trees and shrubs help in attenuating sound pollution besides absorbing air pollutants and beautifying the area. • Listing of sound absorbing materials and composing slogans informing the public about the hazards of noise. • Reduction of traffic and redesigning of engines to reduce vibrations.
15.5	Air is for everyone	Values and attitude	• Certain sensitive zones like hospitals; educational institutions where the noise level has to be reduced and made pollution free by planting rows of trees to absorb noise pollution.
		Conceptual clarity	• Ozone layer getting depleted due to air pollution.
		Application in daily life	• Disposal of radioactive solid wastes from a reactor poses serious chronic problems, e.g. the deadly after effects of the Hiroshima bombing during the Second World War can still be seen in the human race. • Effect of pollution on buildings and monuments, e.g. How SO_2 released by oil refineries is affecting the Taj Mahal.
		Values and attitudes	• Steps taken to clean the air of their homes, schools and other public places by creating awareness about air pollution, understanding kind of vehicles being used, meeting the traffic department to find ways to reduce exhaust emissions, asking traffic policemen about their health problems, surveying the health related problems of roadside vendors and shopkeepers, removing the use of traditional choolahs, establishing an air pollution tolerance index (APTI), trees like *Gulmohar, Neem, Balkan* and *Shisham*, being planted in polluted areas promoting the use of cycles for public transport and obtaining designs of smokeless choolahs and solar cookers from the department of non-conventional energy and popularizing their use in the community by holding meetings and demonstrations.

(Contd)

Sub-unit No.	*Concept*	*Suggested Content*	
		Area	*Actual Incorporations*
			• Story of a student staying near a factory developing skin problems and other defects in the body.
		Field visit	• Visit to the local hospital and requesting the superintendent to help you to obtain information on the prevalence of respiratory ailments like chronic bronchitis, asthma, running nose, eye irritation and headache in your community.
15.6	Water is essential for life	Activity	• Taking details of the air emission control devices from the local pollution control board office and asking them as to whether they have air emission control devices such as multiclones, bag filters, electrostatic precipitators or wet scrubbers. • Experiment on finding out the quantity of dust in the air and as to what will happen to the plants if their tender leaves get covered with dust and further what will happen to animals if they eat the leaves covered with dust and ashes coming from vehicles and factory chimneys.
		Conceptual Clarity	• Effect on the body by swimming in a polluted pond.
		Skill and Processes	• Testing the local water resource water for pollution. • Monitoring pH of water and water testing.
		Field visit/ Assignment	• Visit to the local hospital to find out about the prevalence of water-borne diseases as 60 per cent of all diseases in India are due to contaminated water. Also visit a veterinary hospital to know whether the pollution has affected the cattle. • Visit factories, which are discharging liquid affluents, and ask them whether they have obtained consent from the Pollution Control Board for discharging their affluents.
		Activities/ Assignment	• Monitoring of pollution by listing the determinants prescribed by Central Pollution Board like colour, odour, turbidity, presence of dissolved solids, pH,

(Contd.)

Sub-unit No.	*Concept*	*Suggested Content*	
		Area	*Actual Incorporations*
			presence of dissolved gases, temperature, plankton test, hardness, chloride, dissolved oxygen and microorganisms. • Finding out the nature and location of water bodies, water budgeting, activities causing pollution to water bodies, collecting data on various water-borne diseases in the village discussing methods of recycling water, recycling water used in homes or at community wells and making people aware about drinking water specifications. • Field activity for collection of samples of river water according to a regular schedule or timetable. Observations of the environment around the waters before collecting the samples should be recorded. • Preparing an action plan for controlling water pollution like making a collective representation if the factory or polluting source is not conforming to standards, asking the Municipal Corporation for their plans for constructing puce drains and sewage treatment plants, getting in touch with Sulabh International for making public toilets whose waste can be composted, making posters popularizing, planting trees and shrubs around compost pits.
15.7	Problems created by man-made environment	Values and attitudes	• Discussion of hapless victims of nuclear accidents (like that in Chernobyl in USSR and in Hiroshima and Nagasaki). • Discussions on public properties like bus stops, public toilets, litter boxes, drainage, fences, telephones etc. • Valuing the fact that sanitation starts from an individual and community level.
		Assignment	• Drawing an action plan to clean the neighborhood by designing dumping locations and collection methods, elimination of pollution by substitution of materials and fixing pollutants in another substance, medium or space so that they do not come into contact with humans.

(Contd.)

Sub-unit No.	*Concept*	*Suggested Content*	
		Area	*Actual Incorporations*
		Activity	• Managing waste, i.e. leaves, stems, vegetables, leftover food, papers, old books, polythene bags, old clothes, old shoes, pens, hospital wastes etc. • On waste management like assigning duties for collection of garbage or trash from school, paper management by saving and reusing used paper, skits or simulation games, activities on making paper meche and plastic management. • Making a green house from used polythene, making natural fertilizers from garbage and garden litter. • Differentiating between biodegradable and non-biodegradable items. • Making a map showing approximately the position of the garbage bins and houses in the locality. Interview rag pickers/*kabadi walas* and suggesting methods to solve their problems and designing low cost instruments or giving them hand gloves.
		Values and attitudes	• Water management by finding out extent of water wastage in schools and to take steps to reduce the wastage, finding out the number of leaking taps in the school and repairing them, making posters on the wastage, being vigilant about open taps and channelling used water into garden patches.
		Application in daily life	• Planning sanitation oriented homes, planting trees and herbs, planting trees around urban homes, hydroponics, balcony pots, bottle plants. • Using indigenous methods for keeping the surrounding clean, e.g *Neem* smoke for controlling mosquitoes. • Judicious use of pesticides.

The lesson on pollution had five sub-units with the possibility of STS incorporation. This lesson was more of a theoretical one as there were no activities and field visits which could have been incorporated to the maximum. Activities like celebrating an environment protection day in the school and community, making the area green, monitoring of pollution, planning sanitation and waste management could be included. Field visits could be made to a local hospital, pollution control board or to the transport department to know the extent of various kinds of pollution in the area. Value and attitudes regarding controlling pollution, greening the area, making the air, water and environment safe for your own self and for others need to be included. There was not even a single diagram in the lesson. A diagram showing the various sources of pollution could have helped to clarify the concept of pollution. Conceptual clarity was also needed for the effect of various pollutants on human body.

Table 7.9: Light Refraction (Class VIII)

Sub-unit No.	*Concept*	*Suggested Content*	
		Area	*Actual Incorporations*
1.1	Refraction	Application in daily life	• Effects of refraction of light observed in many day-to-day events. • Where the concept of refraction is used in various instruments of daily life. • The apparent depth of water which is due to refraction. Desert travellers sometimes see mirages which are caused by refraction on hot days. The passengers in a car may see what appear to be wet patches on the road ahead reflecting light from the sky and mysterious ships seen by some sailors are also mirages. • Giving examples of the apparent position of the stars and the sun (because of refraction). • The spearing of fish requires much skill because allowances for refraction must be made.
		Diagrammatic representation	• Of its uses in the daily life should be given.
		Field visit	• Visiting a pond to see the apparent and the actual depth
1.8	Simple microscope	Activity	• Making a simple lens by using a fused bulb and water or a matchbox microscope or by catching a drop of water in a small loop of wire.

(Contd.)

Sub-unit No.	Concept	Suggested Content: Area	Suggested Content: Actual Incorporations
			• Seeing various small objects under a microscope like the veins of a leaf, dirt on the hand, various food items, vegetables. All these activities will help students to see the value of washing hands before eating and eating washed vegetables and fruits. • Diagram to show how a magnifying glass acts as a convex lens. Focusing the sun's rays on a piece of paper so that a small spot of light is formed and observing what happens.
		Application in daily life	• Palmists study various lines on the hand with the help of a simple microscope.
		Skill and processes	• Observation of feathers of various birds in their village can help them in seeing the difference in patterns. • Teaching the correct skill of handling and using the simple microscope. • Observing drops of drinking water under a simple microscope to show them whether water is safe for drinking or not. Water from various sources can also be seen to observe contamination. • Observing an onion peel and their own cheek cells under the simple microscope.
1.9	Compound microscope.	Diagrammatic representation	• Of the compound microscope
		Skill and processes	• Skill of handling and using a microscope should be given. • Skill of mounting a slide using a cover slip, with diagrammatic representation.
		Application in daily life	• Observing microorganisms of diseases like tuberculosis, under the compound microscope. Blood tests to detect various diseases can also be done with the help of compound microscope.
1.10	Telescope	Diagrammatic representation	• Of the telescope
		Skill	• Handling the telescope
		Activity	• Making a telescope in the classroom.
1.11	Photographi c camera.	Activity	• Making a pinhole camera.
		Diagrammatic representation	• Parts of camera.

(Contd.)

Sub-unit No.	*Concept*	*Suggested Content*	
		Area	*Actual Incorporations*
		Skill	• Using a camera by giving practical experiences on focusing and its step-wise procedure with diagrams.
		Conceptual clarity	• Do's and don'ts while taking photographs. • Knowledge about developing photographs.
1.12	Construction of human eye.	Conceptual clarity	• Comparison of pinhole camera with human eye in context of image formation with diagram. • Picture of how iris changes with dim light and bright light.
		Conceptual clarity	• Discussion of noise sources around and the effect of high intensity noise on human being.
		Field visit	• Practical experience of eye examination by visiting a local hospital.
		Skill and processes	• Testing binocular vision by closing an eye and noticing the differences in seeing a picture, judging distances, seeing apparent movement of a pencil and how an object is seen double. • Reading of doctor's prescription about eyesight power.
		Value and attitude	• Importance of eyesight in human life and tips safeguarding it.
1.13	Some defects of eye	Application in daily life	• Awareness about diseases like night blindness, colour blindness, conjunctivites and other vision defects. • Diet which helps in improving eyesight. • Precautions to be taken to protect the eye. • First aid in case of common diseases of eye.
		Activity	• Investigation to detect colour blindness.
1.16	Light spectrum	Conceptual clarity	• Consolidate ideas on white light, spectra and a rainbow.
		Activity	• Construction of Newton's Colour Disc. • Raindrops acting as prism and splitting up white sunlight into seven colours.
1.18	Primary colours and rainbow	Activities	• With colours e.g. creating a rainbow by using three torches each one covered with either red, green or blue cellophene paper. When these three torches are shined on white screen a rainbow is created.

(Contd.)

Sub-unit No.	*Concept*	*Suggested Content*	
		Area	*Actual Incorporations*
			• Testing of different colour combinations (making signboards for the school in different colours and testing possible colour combinations to see which is most effective).
		Skill and processes	• Watching the effect of different colours on the rainbow. • Of making secondary colours from primary colours.
		Value and attitude	• Importance of coiour in nature and in human life.
		Activity	• Stimulating and interesting exercises on road safety like making and using road safety tops, jackets, sashes, belts, reflective strips etc. (brightly coloured ones which are effective in normal light and reflective ones which are particularly effective at night when illuminated, by car lights). • On the idea of functional and interesting clothing suitable for the climatic conditions. • Interesting games with a set of cards with faces drawn in different colours, when seen with different colour filters colours become darker while others disappear. • Seeing traffic lights under sunglasses of different colours.
		Application in daily life	• Knowledge of colours to be used for painting walls and furniture of school and home.

The lesson on light refraction dealt with the theoretical aspects of laws of refraction, ray diagrams of various lenses, prisms and structure of the eye. Activities regarding these concepts and the skill and processes involved in handling various optical instruments was totally lacking. Need was felt to incorporate activities like making the simple lens, pin hole camera, investigation on colour vision, construction of Newtons' disk and stimulating and interesting exercises on road safety. Skill and processes involved in handling simple microscope, compound microscope, telescope, camera, reading eye sight power and making secondary colours need to be incorporated. Application of the concept of refraction in daily life like distinguishing between apparent and real depth of water, mirages, position of stars, use of simple microscope and compound microscope for daily life purposes, awareness of eye defects and its remedies and precautions, application of knowledge of colours for painting etc. need to be dealt with. Diagrammatic representation regarding ray diagrams were included but there were no diagrams for parts of a compound microscope, telescope and camera. Conceptual clarity was required for spectra, handling of a camera and comparison of pin hole camera with human eye. Value and attitude for keeping eyes safe and the importance of colours in nature need to be emphasized.

Table 7.10: Agriculture (Class VIII)

Sub-unit No.	*Concept*	*Suggested Content*	
		Area	*Actual Incorporations*
14.3	Field Preparation	Activity	• The collection of different kinds of soil and their identification. • Practical preparation of nursery soil. • Field preparation for some local crop should be mentioned, like, mustard wheat, paddy, sugarcane and bajra. Activities related to it should be given. • Different ways of ploughing, i.e. from outside to inside and from inside to outside needs a mention.
		Diagrammatic Representation	• Different types of soil. • Mastel and Sabash plough with the uses of all its parts. • Different types of harrow and their uses.
		Application in daily life with diagrammatic representation	• Discussion on different methods of conserving rainwater in the soil, e.g. depth ploughing of fields in summer, bundling in the field, levelling of the soil. Growing of crops in rainy season on slopes. • Conservation of soil moisture by avoiding weeding, hoeing, manuring and allowing the soil to remain fallow. • Discussion on ways of improving the soil for agriculture like improving the acidic and alkaline nature of the soil by making proper arrangements for the outlet of water, by use of proper chemicals, use of potash,

(Contd.)

Sub-unit No.	Concept	Suggested Content: Area	Suggested Content: Actual Incorporations
			the surface of the soil, scratching of the soil, bunding, summer ploughing, leaching of water, change of complex salts to simple salts etc.
		Conceptual clarity	• Protection of grass fields by plantation of trees, which can prevent soil erosion. • Primary tilling depending upon the type of crop, for example, jawar, bajra, and maize require two ploughing, more than two deep ploughing is required for wheat and barley and more than four ploughing for potatoes and sugarcane. • Tilling of land in various seasons like summer, winter and rainy seasons. Listing the advantage of tilling the land in three different seasons. • Knowledge of different types of agricultural instruments used for ploughing based on the type of work required i.e. to cut the soil, dig and turn the soil, to make the soil fine etc.
		Activity	• Practical preparation of nursery soil. • Field preparation for some local crops should be mentioned like wheat, mustard paddy, sugarcane and bajra. Activities related to it should be given. • Different ways of ploughing i.e. from outside to inside and from inside to outside needs a mention.
14.4	Sowing	Value and attitude	• Identification of good quality seeds. • Precautions to be taken at the time of selection of seeds.
		Activity	• Practical classes on sowing different types of seeds.
		Conceptual clarity	• Why a particular method is suitable for a particular type of crop. Merits and demerits of different types of sowing with examples. • Precautions to be taken at the time of sowing like seed treatment for quality, sowing time, quality of seed, early or late variety, time of maturity.

(Contd.)

Sub-unit No.	*Concept*	*Suggested Content*	
		Area	*Actual Incorporations*
		Diagrammatic representation	• Of sowing methods.
		Field visit	• To a seed market for proper identification of seed.
14.5	Weeding	Application in daily life	• Instruments used for weeding like Racket, Spade, and Hand hoe with their diagrams. • Quantity used is important e.g. For stem F34, 8 litres per thousand litres of water per hectare.2, 4D, 600 grams per 600 litre water per hectare. • Harmful effects on the environment, human life and animals of different weedicides commonly used should be explicitly mentioned, e.g. DDT controversy could be discussed in the class.
		Skill and processes	• Proper stepwise use of weedicides with precautions and practical classes.
14.6	Manuring	Diagrammatic representation	• Of different methods of manuring.
		Field visit	• To the fields to show the practical experience of different methods of manuring.
		Conceptual clarity	• Of which manure is good for which crops with examples like cake manure for Mustard and groundnut.
		Skill and processes	• Practical exercises on making manure.
14.7	Irrigation	Values and attitudes	• Importance of irrigation for plants and sources of irrigation like canals, tube wells, wells, ponds, lakes and dirty water to be discussed.
		Diagrammatic representation	• Of various resources of irrigation. • Of different irrigation methods.
		Conceptual clarity	• Machines used for irrigation and their types: (1) manual instruments like bedi and dhakuli (2) machines operated with the help of animals like charsa and rehat (3) electrically or diesel run machines, tube well and water pumps. • The best irrigation methods and procedures for different crops.

(Contd.)

Sub-unit No.	Concept	Suggested Content	
		Area	Actual Incorporations
		Application in daily life	• Methods of irrigations with their merits and demerits. Methods like flooded, beds and burha, furrow and channel and thala method.
		Skills and processes	• Practical exercise on these irrigation methods.
		Field visit	• To the nearest agriculture fields at the time of sowing and irrigation.
14.8	Crop protection	Conceptual clarity	• Types of crop protection from animals and insects and to protect crops from diseases. • Knowledge about companion plants for example with sweet potato, tomato, chilli and beans.
		Skill and processes	• Preparation of home made sprays. • Preparing fire retardant building material. • Devising an electronic fence. • Making eco-friendly pesticides and green fertilizers.

The lesson on agriculture introduced various terms used in the agriculture practices without much clarity regarding their use in actual fields, for example in the sub-unit field preparation, various types of tools used for it were given and the uses of field preparation were given but the procedure for field preparation was totally lacking. Similarly, in the sub-unit sowing, irrigation manuring and crop protection, conceptual clarity was required regarding their procedure. The skill and processes involved in removing weeds, irrigation, manuring and crop protection needs a mention. Real hands on experiences should be given in preparing the field and sowing. Diagrammatic representation was lacking in the lesson. Diagrams of the agricultural practices were needed which would clarify the procedure. Real agricultural fields should be visited at the time of various agricultural practices. Positive values and attitudes need to be developed while buying and selecting various kind of seeds and for proper usage of water from various sources at the time of irrigation.

The incorporation made can be summarized in the form of Table 7.11. The whole summarization portrays that lessons lack diagrammatic representation, activities, examples/application of daily life, skill and processes, values and attitudes, conceptual clarity and lastly assignments and field visits. Diagrammatic representations, skill and processes and examples from daily life had been present in the text to some extent but activities, values and attitudes oriented content was lacking. A special attention needs to be given to these while writing the textbooks. An STS orientation of the text will compel the writer to include all the aspects to be mentioned here which have been averagely neglected.

Table 7.11: Summarization of the Incorporations made in the Six Lessons

Lesson	*Total Sub-units*	*Areas*						
		Diagrammatic Representation	*Activities*	*Assess-ment/ Field Visits*	*Conceptual clarity*	*Examples/ Applications in daily life*	*Skill and Processes*	*Value and attitudes*
Measurement	11	5	7	1	2	7	5	4
Water	8	8	5	2	6	4		5
Population	3	2	2	1	3	3	2	4
Pollution	5	1	5	4	3	3	2	4
Light Refraction	9	4	6	2	3	5	6	2
Agriculture	6	4	2	3	5	3	4	2

Knowledge of science is growing exponentially and one has to decide seriously how much of it should be covered in school textbooks. At the same time, the emphasis of science education has shifted from feeding information to the students as an active participant in the learning process. This would enable students to develop abilities to apply the knowledge and processes of science learned in school to solving problems of daily life. The thrust is on understanding societal issues as they relate to science and technology and being able to make value judgement regarding them. These aspects of science education have to be taken into account while writing textbooks. Thus, there is a need to emphasize a content-process-context based STS curriculum which would present the true image of science. Since the image of science presented in classroom is often limited to the textbooks, there is need for a balance presentation of all aspects of science.

Transaction of STS Oriented Content

STS could have a negative effect on students attitude, specially if STS were to be taught in a "traditional" way, with students largely in a passive mode. Teaching methods and strategies should be considered along side any curriculum change. A teacher of STS who uses a variety of teaching techniques, some of which could be games, simulations, role play and other strategies to make students more active, will tend to create a lively, stimulating classroom climate which will be perceived by students as being more enjoyable. In this way students will have a more positive attitude towards STS. Since STS is taught in a social context and most of the games and simulations are done in social science, these could be said to relate more to STS than it does to science education itself. The linkages between science, technology and society are complex and therefore, require different methods to be adopted in classrooms to encourage students to form such links. These linkages could be developed through suitably designed activities, which include the following teaching strategies:

1. *Stories and Pictures*—Stories told and pictures shown can make the content more interesting. It helps them relate the concepts to their life situations. Students can derive morals, ethics and values related to science through stories and pictures. The story of the two farmers in the 'Water' lesson, is an example. Pupil's strip cartoon also makes learning enjoyable (as given in the Appendix G).

2. *Local Newspaper*—The use of local newspapers is good way to teach local issues. Every locality has difficult issues, which can be solved through teaching of science. Being in a familiar setting, pupils are in a better position to understand the concerns of the people. The teacher may arrange to collect newspaper cuttings on problems affecting the neighborhood over a few months and use them at an appropriate point in her scheme of work. This kind of exercise can also be done when there are news breaks—it may be regarding pollution of the water supply or air, the outbreak of various diseases, controlling population or any other problem.

Pupils can also be asked to collect the information from newspapers, the local radio and from people that they know which is clearly an advantage. By doing this, the issues will be real to them and are about people they know. It would enable them to begin to form their own opinion on a civic issue and even to see what action might follow to deal with it. For example, in the lesson of measurement reading of temperature in the newspaper can be discussed. In the lesson of water valuing the importance of water can be taught by discussing newspaper clippings on water problems faced by the state (as given in the Appendix G).

3. *Simulation and Role Play*—Activities that include simulation and role-play help solve everyday problems using the science that students have learnt in school. Simulation puts pupils in the position of being detectives, engineers, architects, managers etc. This helps to strengthen their skills and acts as an incentive to pupils to become involved in their work. The approach can apply to the presentation of problem to reporting back upon completion of an investigation. Individual group members can take on tasks which suit their individual strengths and preferences. For example, the more confident, articulate members could talk about their work, while others could prepare diagrams, models or written texts to suit their preferred skills. Young pupils can often produce excellent posters describing and illustrating what they have done, sometimes with comic inventiveness. A display of these posters on the walls of the laboratory will stimulate interest in other classes. Many of these activities are time consuming. Nonetheless time should be taken at the upper primary level to develop these skills which will be very useful to pupils as they progress through their secondary education.

In role-play, pupils take on parts and act them out in order to feel the likely reactions of the characters, as they perceive them. The pupil can often bring the character to life, extend his sympathy and so begin to think about the causes of mistreatment and the means of curing the abuse. Role-play fulfils the essential aims of showing that science is about people. For example, in the lesson of population the concept of a small family and large family can be taught through role play and simulations.

4. *Games*—Games are an effective teaching strategy. A game is a lighthearted simulation; it not only amuses students but also encourages them to discuss issues related to their lives. Games in the form of open-ended questions such as the consequences in terms of pollution, noise and disruption of the local community can be put to a group. Games like snakes and ladders, dice, card games about subsistence farming, malnutrition, life threatening diseases and the poverty game now renamed for the farming game can be played.

5. ***Quiz and Asking Provocative Questions***—Quiz helps the teacher focus on topics less understood by students and which need to be supplemented with facts, exercises and additional activities. For instance, when dealing with large numbers the students are often unclear about conversions between lakhs, crores, millions and billions—Indian and international place value systems. By asking provocative questions commonly referred to as open ended or extended answer questions teachers can stimulate the critical thinking skills and get children mentally excited about inquiry, exploration and the discovery of new knowledge. For example in the lesson of light refraction asking the old age question as "why is the sky blue?"... Can be alternatively asked through questions like "why isn't the sky red?"... How could you make the sky red?... Why at times, does the sky appear to be red?... Or how would a different coloured sky affect your environment?...

6. ***Providing Paradoxical Situations***—A paradox is a true situation that seems contradictory, unbelievable or absurd: but may in fact be true. It is an observation that is inconsistent with common experience. For example, in the lesson of water the interaction of wood and water as an activity is an example. After students generate a list of expected outcomes about the interaction they set about testing their hypotheses. The introduction of lignin vita, a type of wood with a specific gravity greater than water, provides a discrepant event as it sinks and creates a contradiction between many expected outcomes and actual observations.

7. ***Teaching Role of Habits***—Creativity can be enhanced by developing sensitivity to the effect of habit bound thinking, thus allowing students to examine the advantages and disadvantages of convergent *versus* divergent thinking. For example, in the lesson of agriculture, a simple weeding instrument like a racket can be shown to the students and they could be asked to generate a list of uses. After discussing their responses, continue the collection of ideas in a less habit bound and more divergent setting.

8. ***Problem Solving***—Focus on problem identification not merely on problem solving. This strategy provides the student with the opportunity to grapple with the difficult task of identifying a problem and the process involved in searching for casualties. For example, students design their own experiment by identifying a researchable problem, describing a logical path of investigation and formulating expected or potential outcomes. The last step is the testing of their hypothesis and reporting. Problem solving methods provide situations that allow greater student control over variables and hypothesis formulations. They create open ended situations that do not force closure of final answers. They leave the door open for the students to continue thinking, developing and testing long after the formal class activity has ended. This strategy forces the student into situations where there is no simple correct answer. It more accurately portrays the nature of science, that ultimate answers do not always exist. For example, in the lesson of light refraction, students build an eye testing device. This strategy can be combined with the open ended questions like... How do you improve the testing device?... Give an explanation for how it works etc....

9. ***Developing Skill in Creative Writing and Reading***—With this approach students are taught to express their ideas and feelings clearly and concisely. In the same manner, students are taught to read for meaning and to search for implications. For example, students examine newspapers, television and magazines for statements and paradoxical situations related to various causes of pollution. The teacher helps them to differentiate between variable information and assumptions or editorial assertions. Also students develop a tentative position on a science related issue report and examine data relative to the issue, and express their conclusions and implications based on their assessment.

10. ***Self-Evaluation***—Encourage self-evaluation whenever possible the student must be given the opportunity to self-evaluate and sense the intrinsic value of learning. This strategy helps students dare to be themselves by providing a classroom environment free from pre-mature external evaluation by the teacher. For example, in the lesson of pollution, the students may be asked to collect water from different sources of water separately. They are asked to check the water for pollution independently and not present the results with reasons in front of the whole class. Such assignments enhance student creativity and allow them to assess the final results. After a class presentation of findings, students assess the degree of task completion by responding to several questions.

The various activities involved for the transaction of these strategies need careful planning by the teacher. The activities to be designed have to be as flexible as possible. They should be fit for a number of different teaching strategies and various approaches should be used which will give much of the initiative back to the pupil, which is investigative, leading to the acquisition and development of a range of process skills. The teacher can ask the students to discuss, research and propose solutions to the problem. Students can plan, carry out an investigation to solve the problem, set, design, build and test a device to meet the criteria given in the problem. The activities should be done in a participatory and interactive mode; sessions should involve thinking across disciplines, stretching the ability of participants to think beyond the obvious connections.

Relevance of issues to daily life should be stressed and participants should be guided in making decisions. During the activities, weaknesses and lacunae should be assessed at intervals through appropriate questionnaires.

One factor, which is common, to all the teaching strategies are that pupils work in small groups of about three. This system of working and sharing ideas within a small group removes much of the personal pressure which some pupils might feel when first presented with a problem. The contribution of individuals with their own strengths can allow the team to work quickly and logically and gives less confident pupils the opportunity to take part. There are pupils who when given a difficult problem, are able to solve it methodically and logically with no assistance from the teacher. However, most groups will need guidance and it is gauging of when and where to give the advice, clue or hint, which is so crucial. To give too little help could leave a group struggling and disheartened and remove any confidence the pupils might have in solving future problems. To give too much help could take away much of the satisfaction pupils might feel upon successfully solving a problem. The best type of clue is where pupils use readily available resources like the library, magazines, photographs, posters, diagrams and models to research possible solutions. The arrangement of the classroom should allow for maximum number of students to access the resources. Upon entering the classroom, students should find a wide range of objects available and class time should be provided for students to explore with them. This stimulates student's interest and establishes mental structures before presenting an underlying concept.

Impact of STS Input on USTS and Achievement in Science

The modified lessons prepared for class VII were taught by the researcher in the Government High School, Bharawas. Twenty lesson plans were prepared for teaching the topics population and pollution ten for the control group and ten for the experimental group. The researcher taught the lessons both to the experimental and control group. The two sections of class VII were randomly assigned as control and experimental groups.

The effectiveness of exemplar lessons was evaluated in terms of achievement in science and Understanding of Science, Technology and Society (USTS). The assessments efforts revealed some exciting information concerning success with the STS oriented modified lessons. The researcher taught one section of students with STS example material and another from the regular textbook. Pre-test and General Mental Ability test were administered to both sections at the beginning and post-test and USTS were administered at the end of the experiment. To control the differential effect of intelligence and previous knowledge on achievement (post-test) and USTS ANCOVA design was adopted.

The mean of the results as seen in Table 7.12 clearly shows that students of the experimental group are far more successful than the control group. In the post-test the mean of the experimental group was found to be 19.4 (64.7%) which is higher than the mean of control group, i.e. 17.4 (58%). Similarly, for the USTS the mean of the experimental group was found to be 25.8 (51.6%) which was much higher than the mean of the control group i.e. 18.8 (37.6%). Summary of analyses of covariance for post-test keeping intelligence scores and pre-test scores as covariance and for USTS keeping intelligence course and pretest scores as covariates are shown in Tables 7.13 and 7.14.

Table 7.12: Percentage Means of the Control Group and Experimental Group for Post-test and USTS

Test	*No. of Items in the Test*	*Mean of Experimental Group*	*Percentage Mean of Experimental Group*	*Mean of Control Group*	*Percentage Mean of Control Group*
Post-test	30	19.4	64.7	17.4	58.0
USTS	50	25.8	51.6	18.8	37.6

Table 7.13: Summary of Analysis of Covariance for Post-test Keeping Intelligence Scores and Pre-test Scores as Covariates

Source of Variation	*Sum of Squares*	*DF*	*Mean Squares*	*F*	*Significance of F*
Main Effects	74.40	1	74.40	6.52	0.013
Covariates	34.90	2	17.40	1.53	0.225
Pre-test	14.20	1	14.20	1.24	0.269
Intelligence test	25.20	1	25.20	2.21	0.142
Residual	753.71	66	11.42		

Table 7.14: Summary of Analysis of Covariance for USTS Keeping Intelligence Scores and Pre-test Scores as Covariates

Source of Variation	*Sum of Squares*	*DF*	*Mean Squares*	*F*	*Significance of F*
Main Effects	813.87	1	813.87	148.39	0.000
Covariates	45.22	2	22.61	4.12	0.021
Pre-test	16.27	1	16.27	2.97	0.090
Intelligence test	22.79	1	22.79	4.16	0.046
Residual	361.995	66	5.485		

Since there were only two groups in the present experimental design, the value of 't' will be √ F and both F and t became significant at the 0.01/0.05 level. This reveals that the mean performance of the experimental group is better than that of the control group on post-test and USTS when the differential effect of two variables namely pre-test and Group Mental Ability test were adjusted.

The F and t values for the achievement test of the two groups were found to be F = 6.52 and t = 2.55, significant at 0.05 level of confidence. This shows that the group receiving STS input achieved significantly higher than the other group that did not receive any STS input. The F and t value for the USTS of the two groups were found to be F = 148.39 and t = 12.18, significant at 0.01 level of confidence. This shows that the experimental group has significantly higher understanding of science, technology and society (USTS) in comparison to the control group.

Based on the above analyses the two hypotheses formulated at the beginning of the experiment were rejected, i.e. There is no significant difference between the experimental and the control group with regards to the understanding of science, technology and society when differential effect of intelligence and pre-test was adjusted by ANCOVA. This hypothesis was rejected at 0.01 level of confidence.

There is no significant difference between the experimental and the control group with regards to the achievement when the differential effect of intelligence and pre-test was adjusted by ANCOVA. This hypothesis was rejected as 0.05 level of confidence.

Hence, there is a significant difference in the experimental and the control group with regard to the understanding of science, technology and society and achievement. This difference is attributed to the STS input provided in the content which was taught to the experimental group.

Thus, the students taught with STS orientation show higher achievement as their concept formation is greater; they can apply knowledge to new situations. They show better understanding of science, technology and society interactions, connect science and technology to other human endeavors, offer explanation of natural phenomena and understand the social construction of scientific knowledge and technology (USTS domain). Creative skills are enhanced enormously and significantly in experimental group over the control group.

The following analysis is in consonance with the various studies like Yager *et al.* (1990) where three hundred students were tested from grades four to nine of Iowa classroom. One group experienced traditional science and another experienced science with STS focus. Pre-test and post-test was given to both the groups and it was found that students exposed to an STS experience were superior to students in traditional science courses. Iskandar (1991) conducted a similar study which showed that STS students were as capable as textbook-oriented students in mastering basic concepts. They had a more positive attitude towards science, science teachers and science related careers and perceived the power of problem solving significantly better. Similar results were obtained by studies carried out by Yoshisuke (1993), Yager and Tamir (1993) and Kortland (1996).

8

Findings and Implications of the Study

Everyone in the present society needs to have knowledge of science. Science plays an important part in the task of preparing citizens and inculcates various values like discipline, utilitarian vocation and culture etc. Various commissions and committees appointed in India to study and review the problem of education have also unanimously suggested science education as a compulsory part of general education. Earlier, stress was only on disciplinary content of science but today science is so pervasive that interlinkage between science, technology and society determines the social, economical and political structure of any nation.

Science education curriculum should incorporate and emphasize the inter-relationships of science, technology and society to produce educated laymen regardless of the diverse nature of their future careers. Appropriate technology taught to students can have beneficial effects on relieving poverty in a rural Indian environment. The major objectives of rural development, i.e. removal of poverty and unemployment, can only be achieved if the rural population have a positive attitude and acceptance towards the appropriate application of technological solutions. This technology education should be appropriate to the local requirements and technological achievements' impact on everyday life should be integrated in the science curriculum. Pupil should be assisted in understanding the consequences of technology in society, both positive and negative, which will raise the level of scientific knowledge, skills and attitude of the students to allow them to be more productive in their home environment and to the society at large.

Recent developments in science education have emphasized on socially and technologically oriented science curriculum, that is, STS curriculum. STS is the term coined by science educationists, which reflects the useful multiple connections between the terms Science, Technology and Society. It presents the structure of science in relation to society and emphasizes troublesome areas, which are subjective and value laden (Fullick, 1992). Current literature discussed in the earlier chapter is widely supportive of the inclusion of STS curricula in science education. STS programmes and themes have been developed in an effort to interpret science and technology as complex socially embedded enterprise and to promote the development of a critical, scientifically and technologically literate citizens, capable of understanding STS issues, empowered to make informed and responsible decisions and able to act upon those decisions.

Issues of STS curriculum development, its adoption and implementation are the concern of science educators. Action research is a coherent and viable way of addressing these issues. Reforms through research in the curricular process give primary attentions to the development of instructional material, particularly the textbook. Many studies carried out have revealed that the textbooks determine the order, the examples and the application of science topics that students experience. Investigations of the science classroom transactions revealed that 90 per cent of the teachers use a textbook 95 per cent of the time. Since the image of science presented in classrooms is often limited to the textbooks, there is need for a balanced presentation of all aspects of science. A textbook, which emphasizes the knowledge aspect, would indicate to the students that science is an organized body of information. A text, which includes the processes, would convey the investigative nature of science and a text with the inclusion of science, technology and society would convey the immediate interest and the relevance of science in the lives of the students.

The STS model in science education provides a structure for inclusion of all these aspects to be set in a personal, local, national or global context. The orientation of these aspects becomes the building block of the curriculum. The STS model can be used in conjunction with the existing science curriculum to explore issues that are socially relevant and personally compelling or it can be used to formulate separate STS courses. Rather than reconstructing the entire science curriculum around "Social issues" (which may not be feasible in some cases), STS can be integrated into appropriate units in the curriculum.

The purpose of this study has been to look into the possibility of STS incorporation in the already existing science textbooks. For the purpose of STS incorporation, STS themes in science education, which have already been formulated by science educators abroad, were validated in the Indian context at the upper primary stage. The upper primary level was particularly chosen, because the curriculum at this level needs to develop a more utilitarian focus. The curriculum of science education at this stage has been initiated with the objective of producing experts, middle order workers and scientifically literate citizens. The last would include those students who would leave school after the upper primary level. The most important factor in creating scientific literacy would be to make science relevant to the immediate environs and needs of these students. It is probably sensible to begin by gradually adding relevant technological and societal content to existing courses; since this strategy can be implemented relatively, quickly and with less expense, a major requisite of the syllabus, it is less threatening even for teachers. An STS orientation of the content of Haryana science textbooks of classes VI to VIII was carried out and the exemplified lessons were tested in the real classroom situation for their feasibility.

Haryana State was particularly chosen for the study as the state is a representative of the rural backward villages where scientific literate people are needed to overcome the age-old superstitions and practices. The state has made rapid developments in infrastructure and high agricultural growth but despite these progresses, the state continues to register poor performance on all well-established educational and social indicators.

Findings

Analysis of the data by the researcher resulted in the following findings:

(i) The STS theme framework in science education at upper primary stage, validated by the experts consisted of thirty-eight aspects in the various domains, namely acquisition of knowledge, skill and processes, ethics, values and attitudes and application.

(ii) The domain, acquisition of knowledge consisted of three sub-domains, i.e. concepts related to conventional content of science, concepts related to science and technology and concepts related to science, technology and society. There were twelve aspects validated under this domain with a break up of four under each sub-headings, the sub-heading concepts related to conventional content of science, four related to science and technology and another four related to science, technology and society (facts, information, concepts, laws, interaction and change, growth and cycle, patterns and processes, probability and prediction, describe interaction of science, technology and society, understand the limits of science and technology, distinguish personal opinion from scientific evidence and understand application of technology).

(iii) In the second domain, skill and processes, eleven aspects were validated, i.e. questioning and searching, observing using space time relationship, classifying, grouping and organizing, interpreting data, comparing and conserving, critically thinking skills, use of technology in practical situations, producing mental images, combining ideas in new ways, predicting possible consequences, suggesting possible causes of given situation.

(iv) In the third domain, ethics, values and attitudes, seven aspects were validated, i.e. valuing democracy and democratic process, identifying values bearing on any given STS issue like local issues, public policies, global problems, developing positive attitude towards science in general, science in school and science teacher, developing positive attitude towards oneself, expressing personal feeling in a constructive way, developing sensitivity to and respect for the feelings of other people, making decisions about social environment issues and exploring arguments on various sides of an issue.

(v) In the fourth domain, seven aspects were validated, i.e. seeing instances of scientific concepts in everyday life experiences, understanding science and technology principle involved in household devices, using scientific process in solving problems in everyday life, making decisions related to personal health, nutrition and lifestyle based on knowledge of scientific concepts, extended school experience beyond classroom, handling of apparatus and development of low cost materials.

(vi) During the analysis for general presentation and layout of the textbook, following observations were noted: language of the book was tough as difficult Hindi words were used to explain the concepts. The illustrations present in the textbook were big in size but were not self-explanatory, colourful and labelled. There was not even one activity per lesson given in any of three textbooks. Even where activities were given, they were mentioned under home assignment. There was no categorization of the activities into students' activities and teacher's demonstration. The questions given at the end of the lessons were satisfactory. There were short, long and objective type questions at the end of the lessons. There was no supplementary material available to the students and no teacher's handbook was provided to teachers to assist them in their teaching.

(vii) Class VI textbook gave maximum weightage to the domain, acquisition of knowledge, i.e. 44.7 per cent and within it also maximum importance was given to the knowledge related to conventional content of science, i.e. 28.5 per cent and then to knowledge related to science and the technology, i.e. 13.2 per cent, and lastly, to knowledge related to science, technology and society, i.e. 3 per cent. Skill and processes had a 31.7 per cent weightage. The weightage given to application domain was 22.2 per cent and to the domain, ethics, values and attitudes it was 1.7 per cent.

(viii) Class VII textbook gave maximum weightage to the domain, acquisition of knowledge, i.e. 42 per cent then skill and processes, i.e. 27.4 per cent, then application, i.e. 27.8 per cent and least importance was given to ethics, values and attitude, i.e. 3 per cent. Within the domain acquisition of knowledge, maximum coverage was for the content related to the conventional content of science, i.e. 24.8 per cent, then to science and technology, i.e. 13.6 per cent and least to the knowledge related to science, technology and society interaction, i.e. 3.6 per cent.

(ix) Class VIII textbook gave maximum weightage to the domain, acquisition of knowledge, i.e. 42.5 per cent. The domain application area comes next, i.e. 31.9 per cent. Skill and processes had a depiction of 22 per cent and least presentation, i.e. 3.7 per cent, was of ethics, values and attitudes. Within the domain, acquisition of knowledge, major thrust of the content was on knowledge related to conventional content of science, i.e. 24.4 per cent, then to knowledge of concepts related to science and technology, i.e. 11.3 per cent and least for knowledge of concepts related to science, technology and society, i.e. 6.8 per cent.

(x) Acquisition of knowledge domain covered the maximum content, i.e. 44.7 per cent in class VI, 42 per cent in class VII and 42.5 per cent in class VIII. Within the knowledge, concepts related to conventional content of science was 28.5 per cent for class VI, 28.4 per cent for class VII and 24.4 per cent for class VIII. The conventional content of science should be low for lower classes but it was contrary. The coverage of concepts related to science and technology was 13.2 per cent for class VI, 13.6 per cent for class VII and 11.3 per cent for class VIII. More technological concepts can be taught at higher stages, therefore, the ideal situation would have been when this variation increased with class level instead of decreasing, as shown by variation from class VII to VIII. The concepts related to science, technology and society had least converge, that is 2.9 per cent for class VI, 3.5 per cent for class VII and 6.6 per cent for class VIII. The variation within this sub-category is positive, as these concepts should have more presentation with the increasing class level.

(xi) In the domain, skill and processes, the percentage of the text that deals with skill and processes is 31.7 per cent for class VI, 27.4 per cent for class VII and 22 per cent for class VIII. Skill and processes have an average depiction in the content and its representation decreases from classes VI to VIII.

(xii) In the domain, ethics, values and attitudes, the percentage of text that deals with it is 1.64 per cent in class VI, 3.6 per cent in class VII and 2.84 per cent in class VIII. There is a rise from classes VI to VIII but it is a very meager rise. Some of the aspects under this domain like valuing democracy and democratic process, developing positive attitude towards science in general, science in school and science teacher, developing positive attitudes towards oneself, expressing personal feelings in a constructive way, developing sensitivity to and respect for the feelings of other people were either totally missing or had been negligibly represented in the text.

(xiii) For the domain, application, the percentage of text that deals with it is 21.4 per cent in class VI, 26.7 per cent in class

VII and 31 per cent in class VIII. There is a rise in its depiction from classes VI to VIII, which is a positive aspects but some aspects like development of low cost materials, further extending school experience beyond classrooms, handling of apparatus, understanding science and technology principles involved in household devices and making decision related to personal health, nutrition and life style based on knowledge of scientific concepts are totally missing or have been poorly depicted in the content.

(xiv) The possibility of STS incorporation was present in all the lessons of the class VI textbook and it varied from 73 per cent to 11 per cent. The lessons, namely, Measurement and Water, had the maximum possibility of incorporation of STS in the content. Though the lessons having sub-units with STS themes cover the application domain of the conceptual framework, the skill and ethics, values and attitude domain is missing.

(xv) The possibility of STS incorporation was present in all the lessons of class VII textbooks and it varied from 75 per cent to 13 per cent. The lessons, namely, Population and Pollution, had the maximum possibility of incorporation of STS in the content. The lessons which already had the STS element like Transmission of Heat, Diseases and their Control and Animal Care had scope for further modification by creatively formulating activities which would develop skill areas, processes, ethics, values and positive attitudes. Some other lessons like Pressure and Buoyancy, Nature and Composition of Substances and Your Own Protection had a major thrust for knowledge of higher learning. Some of the concepts in these lessons are difficult to be taught at standard VII level. These concepts should be dealt with in higher classes or alternatively these concepts should have been given in simpler form using different innovative strategies.

(xvi) The possibility of STS incorporation was present in all the lessons of the class VIII textbook and it varied from 50 per cent to 5 per cent. The lessons Light Refraction and Agriculture had maximum possibility of STS incorporation. This percentage is lesser than the percentages of class VI and VII. This is probably due to the reason that some lessons like Electrical Energy, Carbon and its Uses, Acid, Bases and Salts, Crop Cultivation, Useful Plants and Animals and

Science in the Use of Man, had a fair incorporation of STS in their content, but they lacked practical exercises to be done by the students. There are some lessons like Potential Difference and Resistance, Electromagnets, Composition of Atom, Cell and Tissue and Liquids where the content deals with concepts of higher learning.

(xvii) Modifications of the chapters showed that maximum incorporations were made in the area of activities, values and attitudes, diagrammatic representations skill and processes and application/examples from daily life. In some sub-units conceptual clarity and assignments/field visits also needed to be incorporated.

(xviii) Mean performance of students exposed to STS input on post-test was found to be 19.4 (64.7%) whereas the mean performance of students of control group was found to be 17.4 (58%).

(xix) Mean performance of students exposed to STS input on USTS was found to be 25.8 (51.6%) whereas the mean performance of students of control group was found to be 18.8 (37.6%).

(xx) When the differential effect of two variables, i.e. General Mental Ability and pre-test were adjusted, the performance of the experimental group on posttest was found significantly better than that of the control group ($F = 6.52$ and $t = 2.55$, significant at 0.05 level of confidence).

(xxi) When the differential effect of two variables, i.e. General Mental Ability and pretest were adjusted, the performance of experimental group on USTS was found significantly better than that of the control group ($F = 148.39$ and $t = 12.18$, significant at 0.01 level of confidence).

Conclusion

On the basis of the findings the researcher could draw the following conclusions:

1. STS framework for science education of upper primary stage is rated as important for giving a new direction to the already existing Indian science curriculum. This framework covers primary areas, like acquisition of knowledge, skill and

processes, ethics, values and attitudes and application, that were rated as important areas in which the objectives with STS orientation can be formulated.

2. While writing the textbooks the authors and publishers should keep in mind the following aspects:
   - The paper used for the textbook should be soft and white and the printing of the matter should not be smudged. Binding of the book should be strong so that students can use it for at least one year.
   - Language being the most important tool for communication, it should be simple and comprehensible.
   - The illustrations should be big, clear, labelled and colourful as the illustrations attract students to read the content and make the content interesting.
   - Many hands on and value based activities should be included.
   - Student activities and teacher demonstration should be clearly mentioned in the textbook.
   - The main points, which help the students in revision and the teacher to design and prepare for class, should be given at the end of the chapter.
   - Various processes and skills involved in different activities should be given with the diagram at the end of the textbook.
   - Supplementary reading material and teacher's handbook should also be provided for better transaction of the content.
3. The text of all the three textbooks of classes VI to VIII were laden more with theoretical acquisition of knowledge rather than with activities, skill and processes, and ethics, values and attitudes. These two areas should be inculcated to the maximum from as early a stage as possible.
4. Within the domain, acquisition of knowledge, more emphasis should be on the concepts related to science, technology and society and this emphasis should increase from lower class level to a higher class level, i.e. from class VI to class VIII. There is need for the content of the textbooks to be more in a social content with a balance of science and technology and with an emphasis on inter-relationships between and among them, values and society at large.

5. Skill and processes have a poor representation in the textbooks. More activities relevant to the immediate environment and needs of the students should be included. Equal depiction should be there in all the class levels, only the complexity should change from lower level to higher level.
6. Ethics, values and attitudes do not find any place in the present textbooks. Attention needs to be focused on the effects of science and technology upon values, ethics and attitudes. Textbooks should assist the individual to build a coherent system of values to survive in the modern technological world.
7. Application domain is averagely covered in textbooks but there is need to give instances with both the positive and negative effects of science and technology.
8. The content of all the lessons in the textbooks has the possibility of STS incorporation. Rather than reconstructing the whole curriculum, the most probable course is to give STS inputs into already existing content.
9. The content with STS orientation can be made more useful by including activities which will develop technological skills and positive values and attitudes.
10. The lessons which cover the objectives other than STS should be presented with interesting examples from day to day life or with comic representations or with different representations to make the complex concepts simpler.
11. Incorporations in the already existing content in Indian text books can be successfully done by incorporating activities, diagrammatic representation, skill and processes, examples from daily life, values and attitude, and field visits with STS orientation for the already existing concepts in the content.
12. The STS input given to the experimental group resulted in higher achievements of the students in comparison with the students of the control group who were taught by the traditional content. Thus STS input results in higher achievement of students.
13. The STS input given to the experimental group resulted in better understanding of science, technology and society in comparison to the control group taught with the traditional content.
14. Various strategies like role-play, simulation, games, story telling, field visits and activities resulted in higher

achievement and better understanding of science, technology and society.

15. Haryana Education Board textbooks of upper primary stage need improvement in terms of giving emphasis to the STS themes. There should be emphasis on social issues that relate to science and technology.
16. Infusion of STS themes in the already existing science content has been a successful approach for introducing STS concept in the Indian situation.

Educational Implications

The present study has revealed that Haryana Education Board textbooks of science at the upper primary stage have not covered the STS themes in science education and lack STS orientation. The STS themes were highlighted in the textbooks only to some extent. Content should include all the aspects in the different domains of STS theme framework in science education.

- (i) The textbooks should have STS topics or make the already existing content more socially and technologically oriented.
- (ii) There should be more stress on the ethics, values and attitudes domain in the content.
- (iii) The textbooks should develop an understanding of the nature of science, technology and society.
- (iv) The findings of the present study reveal that achievement and understanding the nature of science, technology and society are enhanced with the teaching of STS oriented content. The approach can also be used for other stages of school education.
- (v) The finding of a significant gain in student achievement using STS input versus non-STS counterparts certainly points towards promoting the implementation of STS instruction within the normal school curricula in India. The curriculum material must be revised and all efforts should be made to make them more activity oriented and child centred.
- (vi) Besides modification in curriculum and the approach of writing textbooks, it is also important that teachers be oriented to deliver the content through activity/discovery approach. For this, regular in service courses on methodology should be conducted.

(vii) If STS topics are to be incorporated into the present or future curriculums, plans should be made to implement in service courses for teachers responsible for teaching the course. Presentation of in service programmes should be done from time to time to update science teachers' content background related to STS themes and issues.

(viii) Researches need to be conducted on the weightage assigned to different domains of science education during the writing of textbooks and the weightage of time to be spent on this domain during the transaction of the content.

(ix) Workshops should be organised in which instruction strategies are developed around the STS themes and also for reviewing existing curriculum material and to develop new materials to meet the individual needs of teachers and school districts.

(x) In the present study STS theme framework in science education was formulated for the upper primary stage. A similar attempt can also be done for secondary and higher secondary stages of education.

(xi) Infusion of STS themes in already existing content was done only for two lessons each of classes VI, VII and VIII. The infusion work can be carried out for all the lessons of the science textbooks.

(xii) The findings imply that in science education teachers should transact the text material with the help of the students in the light of the interactions of science, technology and society, i.e. science should be taught in the social context. The teacher should discuss the importance and relevance of science and the students should be encouraged to adopt science-related hobbies and to organise science clubs.

(xiii) There should be creation of STS resource information centres for science teachers and science supervisors, which could provide valuable information about various STS resources and could also assist curriculum reforms.

Suggestions for Further Research

In the present study an attempt has been made to validate a STS themes in the Indian context and to evaluate the science textbooks of upper primary stage in terms of the validated STS theme framework in science education. Incorporations were also done in a few lessons of all the three classes (VI to VIII). But after completion of the study the investigator felt that the field of research chosen is extremely vast and there remains a great deal to be explored in the area. The "STS curriculum" is itself a very broad term, having different models and different constraints in its implementation, which is a fertile area for research. A lot has been done in this field abroad but not much has been done in this area in India. In the Indian situation this area of science education is very raw and needs a thorough investigation before being implemented in the real classroom situation. Researches could be conducted in the areas given below:

(i) The present study did not include independently the effectiveness of STS in terms of different outcomes/objectives of science education. The effectiveness was seen only in terms of achievement and understanding the nature of science, technology and society. Studies need to be conducted to see the effectiveness of STS on various other domains.

(ii) A longitudinal study is required to find out the impact of traditional and STS oriented curriculum on development of the attitude of students towards science, technology and society.

(iii) Additional research needs to be done to document the nature of children's conceptual changes as a result of the STS learning materials versus more traditional materials and methods of curriculum development. While we possess enough information at present to know for certain that children conceive things in the world differently from adults, we are not aware of the exact nature of their understanding of complex interactions of science, technology and society. Far more is known about children's conception regarding science, however, this is not very pertinent to STS education. Since topics generally focus on the level of simple concepts, and primarily in the physical science, research on students across cultures, geographic settings, STS *versus* non-STS learning material etc. would lead to better understanding of children's views about science, technology and society and enrich our understanding of the complexity of human learning and appropriate curriculum material design.

(iv) Some indepth studies are required to identify the causes of lack of understanding of science, technology and society among students of upper primary and secondary classes.

(v) Studies should be conducted on the difficulties, which could be faced by the teachers in the implementation of the STS curriculum.

(vi) In the present study teacher variable and time effect were not studied. Studies should be conducted to study the effect of these variables on the implementation of the STS curriculum.

(vii) The present study did not go into the evaluation techniques and procedures required for STS curriculum. Studies can be conducted on evaluation procedures and techniques required for better implementation of the STS curriculum. Research and development strategies should be evolved to move forward in more promising directions.

(viii) Further studies should be undertaken to determine the specific needs of rural vs. urban needs in the STS curriculum.

(ix) Research is needed on the scarcity of textbooks and instructional materials in the STS field. Various researches may be carried out on resource material development and their evaluation.

(x) Research can be done on teaching techniques and new strategies for implementation of the STS oriented curriculum.

(xi) A similar research similar to the present one can be carried out for the primary, secondary and senior secondary stages of school education.

(xii) It can be researched whether separate STS courses, preparing specific modules on STS, or infusion in the already existing courses is most feasible in the Indian context.

(xiii) Some tests and instruments need to be developed for judging students', teachers', and science educators' perceptions about STS.

(xiv) An attempt in line of preparing VOSTS (views on science-technology-society) for judging the students' perception about STS can be done in the Indian situation.

(xv) Classroom interactions, decision-making skills etc. can also be studied independently to determine the effectiveness of STS.

(xvi) A large number of STS models were evolved abroad but there is a need to develop such models in the Indian context and also to evaluate their effectiveness.

(xvii) Need is also felt to evaluate textbooks prescribed in different states in terms of the STS themes.

APPENDICES

APPENDIX A

A Comparison of Guidelines and Curricular Frameworks for STS Education

	1	*2*	*3*	*4*	*5*	*6*	*7*	*8*	*9*	*10*	*11*	*12*	*13*	*14*	*15*
Knowledge Area															
1. Describe interaction of science, technology and society	X	X	X	X	X	X	X	X	X	X	X	X	X	X	X
2. Understand limits of science and technology	X		X		X		X				X				X
3. Recognize and/or distinguish scientific from technological developments	X		X		X	X					X	X			X
4. Awareness of current STS issues		X	X	X				X	X			X	X	X	X
5. Awareness of global interdependency				X	X				X				X		X
6. Understand key STS concepts	X	X			X			X	X				X		X
7. Distinguish personal opinion from scientific evidence			X		X	X	X	X	X	X		X		X	X
8. Understand application of technology			X				X		X		X		X	X	X
9. Be able to use correctly some qualification techniques	X	X	X	X						X	X				X
10. Critique proposed solutions to STS issues	X	X			X	X									X
11. Know that STS issues are persistent		X						X	X					X	X
12. Understand some of sociology and epistemology of science and technology			X		X		X	X	X				X		
13. Nature of Inquiry			X						X						
14. Science and Technology are controlled by society			X			X	X	X	X	X	X	X		X	X
Skills Areas															
1. Communications skills regarding STS issues		X	X		X							X		X	X

(Contd.)

(*Contd.*)

	1	*2*	*3*	*4*	*5*	*6*	*7*	*8*	*9*	*10*	*11*	*12*	*13*	*14*	*15*
2. "Science" or cognitive process skills	X	X	X				X	X		X	X			X	X
3. Decision-making and problem-solving skills	X	X	X	X		X	X	X		X	X	X		X	X
4. Critical thinking skills		X	X		X				X	X	X			X	X
5. Group process skills		X													X
6. Group leadership		X													
7. Conflict management and negotiation		X													
8. Use technology in practical situations						X									X
Ethics and Values															
1. Aesthetic appropriation of world, science and technology					X		X	X	X			X			
2. Value democracy and democratic process					X			X	X	X					X
3. Understand importance of values	X	X	X	X	X	X	X	X		X	X	X			X
4. Be able to articulate personal values		X			X					X					X
5. Identify values bearing on any give STS issues				X	X			X			X				X
Action Area															
1. Conduct research	X									X					X
2. Take action on STS issue in local community and/or society at large	X	X		X									X		X

Sources of Statements

1. Rubba and Wiesenmayer (1985, 1987) And Society (1985).
2. Waks (1987a).
3. Aikenhead (1986).
4. Science through Science, Technology 4. 2. Waks (1987a).
5. Northern Examining Association (1988).
6. Science and Technology 11, Ministry of Education, Province of British Columbia (1986).
7. National Science Teachers Association (1982).
8. Hichman, Patrick and Bybee (1987).
9. Science in a Social Context, Addinell and Solomon (1983).
10. National Council for the Social Studies (1983).
11. New York State Education Department (1985).
12. Biological Sciences Curriculum Study (1981).
13. Project Synthesis, Piel (1981).
14. Preparing for Tomorrow's World Project, Iozzi (1987).
15. National Council for the Social Studies (1990a).

Rating Scale for Validation of STS Themes in Science Education at Upper Primary Stage in the Indian Context

Name:

Designation:

Institution:

Education Qualification:

	Essential	*Important*	*Not Important*
I. Acquisition of Knowledge **ज्ञान का अभिग्रहण**			
(a) Concepts related to conventional content of science विज्ञान की रूढ़ विषय सूची संबंधित अवधारणाएं			
(i) Facts तथ्य			
(ii) Information जानकारी			
(iii) Concepts अवधारणाएँ			
(iv) Laws नियम			
(v) Principles सिद्धांत			
(vi) Existing explanation वर्तमान स्पष्टीकरण			
(b) Concepts related to science and technology विज्ञान एवं तकनीकी संबंधी अवधारणाएं			
(i) System and Subsystems तंत्र और उपतंत्र			

(*Contd.*)

(*Contd.*)

	Essential	*Important*	*Not Important*
(ii) Organisation and identity संगठन और पहचान			
(iii) Hierarchy and Diversity सोपानात्मकता और विभिन्नता			
(iv) Interaction and Change परस्पर क्रिया और परिवर्तन			
(v) Growth and cycles वृद्धि और चक्र			
(vi) Understand some sociology and epistemology of science and technology विज्ञान और तकनीक की सामाजिकता और मीमांसा की समझ			
(vii) Nature of scientific inquiry वैज्ञानिक खोज की प्रकृति			
(viii) Science and technology are controlled by society विज्ञान और तकनीकी समाज द्वारा नियंत्रित हैं			
II. Skill Areas and Processes **निपुणता क्षेत्र और प्रक्रियाएँ**			
1. Questioning and searching प्रश्न पूछना और खोज करना			
2. Observing using space/time relationship स्थान/समय संबंध द्वारा अवलोकन करना			
3. Classifying, grouping and organizing वर्गीकरण, समूहीकरण और व्यवस्थापाना			
4. Interpreting data आँकड़ों की व्याख्या			

(*Contd.*)

(*Contd.*)

	Essential	*Important*	*Not Important*
5. Comparing and consuming तुलना और उपभोग			
6. Controlling and identifying variables चरों का नियंत्रण और पहचान			
7. Describing and defining operationally परिचालन द्वारा रचना और परिभाषा			
8. Hypothesizing and predicting परिकल्पना और पूर्वकथन			
9. Communicating skills regarding STS issues STS मुद्दों की जानकारी में निपुणता			
10. Science or cognitive process skills विज्ञान या संज्ञानात्मक प्रक्रिया सुलझाना			
11. Decision making and problem solving निर्णय लेना और समस्या सुलझाना			
12. Critically thinking skills आलोचनात्मक सोच में कुशलताएँ			
13. Group process skills समूह प्रक्रिया की कुशलताएँ			
14. Group leadership समूह का नेतृत्व			
15. Conflict management and negotiating विरोध नियंत्रण और समझौता			
16. Use technology in practical situation तकनीक का व्यावहारिक प्रयोग			

(*Contd.*)

(*Contd.*)

	Essential	*Important*	*Not Important*
17. Developing creativity skills like रचनात्मक कुशलताओं का विकास जैसे:			
– producing mental images मानसिक बिम्बों की उत्पत्ति			
– combining objects and ideas in news ways विचारों और वस्तुओं का नए तरीके से संयोजन			
– producing alternative or unusual uses for objects वस्तुओं का वैकल्पिक और असाधारण प्रयोग			
– solving problems and puzzles समस्याएँ और पहेलियाँ सुलझाना			
– predicting possible consequences संभावित परिणामों का पूर्वकथन			
– designing devices and machines युक्तियों और मशीनों की रचना करना			
– producing unusual ideas असाधारण विचारों की उत्पत्ति			
– suggesting possible causes of a given situation किसी स्थिति के संभावित कारणों को सुझाना			
– Quality and quantity uniqueness of questions प्रश्नों की गुणवत्ता, परिणाम एवं अद्वितीयता			

(*Contd.*)

(*Contd.*)

	Essential	*Important*	*Not Important*
III. Ethics Values and Attitudes **नैतिक मूल्य और दृष्टिकोण**			
1. Aesthetic appreciation of world, science and technology विश्व, विज्ञान और तकनीक का सुरूचिपूर्ण मूल्यांकन			
2. Value democracy and democratic process लोकतंत्र व लोकतांत्रिक प्रक्रियाओं का मूल्य			
3. Understanding importance of values मूल्यों के महत्त्व की समझ			
4. Be able to articulate personal values व्यक्तिगत मूल्यों की स्पष्ट अभिव्यक्ति			
5. Identifying values bearing on any given STS issues like, local issues, public policies, global problems STS मुद्दों के मूल्य की पहचान जैसे- स्थानीय मुद्दें, सार्वजनिक पालिसी, विश्व समस्याएँ			
6. Developing positive attitude towards science in general science in school and science teachers विज्ञान, विद्यालय में विज्ञान और विज्ञान अध्यापकों के प्रति सकारात्मक दृष्टिकोण का विकास			
7. Developing positive attitude towards oneself स्वयं के प्रति सकारात्मक दृष्टिकोण का विकास			

(*Contd.*)

(*Contd.*)

	Essential	*Important*	*Not Important*
8. Exploring human emotions मानवीय भावों की परख			
9. Developing sensitivity to and respect for the feeling of other people दूसरों के भावों के प्रति संवेदना और सम्मान का विकास			
10. Expressing personal feeling in a constructive way व्यक्तिगत भावों की क्रियात्मक अभिव्यक्ति			
11. Making decisions about personal values व्यक्तिगत मूल्यों से संबंधित निर्णय लेना			
12. Making decisions about social environmental issues and exploring arguments on various sides of an issue सामाजिक पर्यावरणीय मुद्दों पर निर्णय और उसके विभिन्न पक्षों की खोज			
IV. Application Area प्रयोग क्षेत्र			
1. Seeing instances of scientific concepts in everyday life experiences वैज्ञानिक अवधारणाओं का दैनिक अनुभवों में दृष्टांत			
2. Understanding science and technology principles involved in household devices घरेलू उपकरणों में प्रयुक्त वैज्ञानिक सिद्धांतों की समझ			
3. Using scientific process solving problems in everyday life दैनिक समस्याओं को सुलझाने में वैज्ञानिक प्रक्रियाओं का प्रयोग			

(*Contd.*)

(*Contd.*)

	Essential	*Important*	*Not Important*
4. Understanding and evaluating mass media reports of science developments वैज्ञानिक प्रगति की रिपोर्ट की समझ और मूल्यांकन			
5. Making decisions related to personal health nutrition and life style based on knowledge of scientific concepts वैज्ञानिक अवधारणाओं पर आधारित व्यक्तिगत स्वास्थ्य और दिनचर्या से संबंधित निर्णय लेना			
6. Integrating science with other subjects विज्ञान का दूसरे विषयों से समायोजन			
7. Taking action to resolve problems or contribute to improvement of local regional, national or international problems स्थानीय, देशीय व अंतर्देशीय समस्याओं को सुलझाना			
8. Involved in community action projects सामुदायिक कार्य परियोजनाओं में भागीदारी			
9. Extending school experiences beyond class room शैक्षिक अनुभवों का कक्षा से बाहर विस्तार			
10. Emphasizing the inter-relationship and inter - connectedness to other human enterprise मानवीय उद्यमों की अन्योन्याश्रितता एवं पारस्परिक संबंध पर बल देना			

(*Contd.*)

(*Contd.*)

	Essential	*Important*	*Not Important*
11. Conducting research and taking action on STS issues STS मुद्दों पर अनुसंधान एवं कार्य करना			

- Mention any other aspect which might be of importance and is missing in this framework.

- Any suggestions would be of immense value for the investigation.

APPENDIX B

Rating Scale for Validation of STS Themes in Science Education at Upper Primary Stage in the Indian Context

Name:
Designation:
Institution:
Educational Qualification:

Domains	*Essential*	*Important*	*Not Important*
I. Acquisition of Knowledge			
(a) Concepts related to conventional content of science			
(i) Facts			
(ii) Information			
(iii) Concepts			
(iv) Laws			
(v) Principles			
(vi) Existing explanation			
Concepts related to science and technology			
(i) System and subsystems			
(ii) Organization and identity			
(iii) Hierarchy and Diversity			
(iv) Interaction and change			
(v) Growth and cycles			
(vi) Patterns and Processes			
(vii) Probability and Prediction			
(viii) Conservation and degradation			
(ix) Adaptation and Limitation			
(x) Equilibrium and sustainability			
Concepts related to Science Technology and Society			
(i) Describe interaction of Science, Technology and Society			
(ii) Understands limits of science and technology			
(iii) Recognize and/or distinguish scientific from technological developments			
(iv) Awareness of global interdependencies			
(v) Awareness of global STS issues			
(vi) Understand key STS concepts			

(*Contd.*)

(*Contd.*)

Domains	*Essential*	*Important*	*Not Important*
(vii) Distinguish personal opinion from scientific evidence			
(viii) Understand application of technology			
(ix) Be able to use correctly some quantification technique			
(x) Critique proposed solutions in STS issues			
(xi) Know that STS issues are persistent			
(xii) Understand some sociology and epistemology of science and technology			
(xiii) Nature of scientific inquiry			
(xiv) Science and technology are controlled by society			
II. Skill and Processes			
(i) Questioning and searching			
(ii) Observing using space/time relationship			
(iii) Classifying, grouping and organizing			
(iv) Interpreting data			
(v) Comparing and consuming			
(vi) Controlling and identifying variables			
(vii) Describing and defining operationally			
(viii) Hypothesizing and predicting			
(ix) Communicating skills regarding STS issues			
(x) Science or cognitive process skills			
(xi) Decision making and problem solving			
(xii) Critically thinking skills			
(xiii) Group process skills			
(xiv) Group leadership			
(xv) Conflict management and negotiating			
(xvi) Use technology in practical situation			
(xvii) Developing Creativity skills like			
– producing mental images			
– combining objects and ideas in new ways			
– producing alternative or unusual uses for objects			
– solving problems and puzzles			
– predicting possible consequences			
– designing devices and machines			
– producing unusual ideas			
– suggesting possible causes of given situation			
– quality and quantity uniqueness of questions			
III. Ethics Values and Attitudes			
(i) Aesthetic appreciation of world, science and technology			

(*Contd.*)

(Contd.)

Domains	*Essential*	*Important*	*Not Important*
(ii) Value democracy and democratic process			
(iii) Understanding importance of values			
(iv) Be able to articulate personal values			
(v) Identifying values bearing on any given STS issues like, local issues, public policies, global problems			
(vi) Developing positive attitude towards science in general, science in school and science teachers			
(vii) Developing positive attitude towards oneself			
(viii) Exploring human emotions			
(ix) Developing sensitivity to and respect for the feeling of other people			
(x) Expressing personal feeling in a constructive way			
(xi) Making decisions about personal values			
(xii) Making decisions about social environmental issues and exploring arguments on various sides of an issue			
IV. Application			
(i) Seeing instances of scientific concepts in everyday life experience			
(ii) Understanding science and technology principles involved in household devices			
(iii) Using scientific process in solving problems in everyday life			
(iv) Understanding and evaluating mass media reports of science developments			
(v) Making decisions related to personal health, nutrition and lifestyle based on knowledge of scientific concepts			
(vi) Integrating science with other subjects			
(vii) Taking action to resolve problems or contribute to improvement of local, regional, national or international problems			
(viii) Involved in community action projects			
(ix) Extending school experiences beyond class room			
(x) Emphasizing the inter-relationship and inter-connectedness to other human enterprise			
(xi) Conducting research and taking action on STS issues			

APPENDIX C

Test on Understanding of Science, Technology and Society (USTS)

विज्ञान प्रौद्योगिकी एवं समाज संबंधित ज्ञान की परीक्षा

कुल 50 प्रश्न हैं 1 प्रत्येक प्रश्न में चार विकल्प हैं, प्रत्येक प्रश्न में सबसे उचित विकल्प पर सही का निशान लगायें।

1. वैज्ञानिकों तथा जादूगरों को तुलना करने के लिये निम्न में से कौन सा वाक्य सबसे अधिक उचित होगा?
   (क) वैज्ञानिक तथा जादूगर दोनों ही वस्तुओं की व्याख्या करने की कोशिश करते हैं।
   (ख) वैज्ञानिक तथा जादूगर दोनों ही वस्तुओं को रहस्यमयी बनाते हैं।
   (ग) जादूगर विषयवस्तु की व्याख्या करने की कोशिश करते हैं जबकि वैज्ञानिक इसे रहस्यमयी बनाते हैं।
   (घ) वैज्ञानिक विषयवस्तु की व्याख्या करने की कोशिश करते हैं जबकि जादूगर इसे रहस्यमयी बनाते हैं।

2. वैज्ञानिक अविष्कार
   (क) विश्व के लगभग सभी देशों में किये गये हैं।
   (ख) केवल बड़े औद्योगिक देशों में किये गये हैं।
   (ग) केवल अधिक जनसंख्या वाले देशों में किये गये हैं।
   (घ) एशिया तथा अफ्रीका के अधिकतर देशों में किये गये हैं।

3. राम ने कहा "वैज्ञानिक का कार्य कभी समाप्त नहीं होता" राम के ऐसा कहने का अर्थ था
   (क) प्रयोगशाला में दिवसीय कार्य-काल बहुत लम्बा होता है।
   (ख) लोग वैज्ञानिकों से बहुत सारे कठिन प्रश्न पूछते हैं।
   (ग) अनुसंधान में वैज्ञानिकों को अत्यधिक सहायता की जरूरत होती है।
   (घ) एक समस्या हल होते ही नये प्रश्न पैदा हो जाते हैं।

4. जब कोई वैज्ञानिक एक अनुसंधान कार्य समाप्त करता है तो प्राय: वह
   (क) अपने निष्कर्ष को अपने देश की सहायता के लिये गुप्त रखेगा।
   (ख) अपने निष्कर्षों और विचारों को अन्य वैज्ञानिकों को जानने देगा।

(ग) उद्योग के लिये होने वाले आविष्कारों में अपने निष्कर्षों को प्रयोग होगा।
(घ) रिपोर्ट लिखने के लिये सरकार से अनुमति मांगेगा।

5. "प्राकृतिक संसार किस प्रकार कार्य करता है" इस विषय में आजकल वैज्ञानिक बहुत से विचारों पर सहमत होते हैं
(क) वैज्ञानिक अधिक जानकारी प्राप्त होने पर बदल जायेंगे।
(ख) आगामी लम्बे समय तक नहीं बदलेंगे।
(ग) शीघ्रगामी संसार की घटनाओं के साथ-साथ बदलते जायेंगे।
(घ) नहीं बदलेंगे क्योंकि वे वैज्ञानिक विचार धारायें हैं।

6. आज भारतीय वैज्ञानिक जो कालेजों और विश्वविद्यालयों में शिक्षण और अनुसंधान का कार्य करते हैं उनकी शिक्षा
(क) प्रायः विश्वविद्यालय की पांच वर्ष की शिक्षा के बाद पूरी होती है।
(ख) प्रायः कुछ उद्योगों में निश्चित समय का प्रायोगिक प्रशिक्षण भी शामिल होने के बाद पूरी होती है।
(ग) प्रायः दो वर्ष की महाविद्यालय अथवा तकनीकी (Technical School) विद्यालय की शिक्षा के बाद पूरी होती है।
(घ) प्रायः महाविद्यालय की शिक्षा समाप्ति के बाद विश्वविद्यालय में एक लम्बे समय का अध्ययन शामिल है।

7. रमेश ने कहा "मशीनें हमारा इतना काम कर रही हैं कि एक दिन ये वैज्ञानिकों का स्थान ले लेगी" रमेश का कथन गलत था क्योंकि
(क) मशीनें अन्य मशीनों का निर्माण नहीं कर सकती।
(ख) मनुष्य संसार को केवल मशीनों पर निर्भर नहीं होने दे सकता।
(ग) मनुष्य मशीनों को स्वचालित नहीं होने दे सकता।
(घ) मशीनें नये विचार नहीं सोच सकती।

8. एक वैज्ञानिक सम्भवतः अपनी छुट्टी के दिन नहीं पसन्द करता है
(क) अपनी प्रयोगशाला जाना।
(ख) अपने मन पसन्द कार्यों में समय बिताना।
(ग) अपने किसी मित्र से मिलने जाना।
(घ) अपने परिवार के साथ आराम करना।

9. आधुनिक वैज्ञानिकों का प्राचीन वैज्ञानिकों की अपेक्षा अधिक कठिन समस्याओं पर कार्य करने का कारण है कि
(क) आधुनिक वैज्ञानिक पहले के वैज्ञानिकों की अपेक्षा अधिक परिश्रम करते हैं।

(ख) आधुनिक वैज्ञानिक प्राचीन वैज्ञानिकों की अपेक्षा अधिक विचार-शक्ति रखते हैं।
(ग) पहले के वैज्ञानिकों द्वारा किये गये कार्य के आधार पर ही आधुनिक वैज्ञानिक आगे कार्य करते हैं।
(घ) आधुनिक वैज्ञानिक प्राचीन वैज्ञानिकों की अपेक्षा अधिक बुद्धिमान होते हैं।

10. आज के महत्वपूर्ण वैज्ञानिक विचार कैसे विकसित होते हैं
(क) एक वैज्ञानिक द्वारा अकेले लम्बे समय तक किये गये अध्ययन के फलस्वरूप।
(ख) बहुत सारे वैज्ञानिकों के कार्यों और विचारों के फलस्वरूप।
(ग) वैज्ञानिकों तथा सरकार द्वारा किये गये संयुक्त कार्यों के फलस्वरूप।
(घ) विज्ञान विशेषज्ञों के इस निर्णय के फलस्वरूप कि क्या अध्ययन करना है।

11. आज सामान्यत: प्रेक्षण (Observations) तथा प्रयोगात्मक कार्यों में वैज्ञानिकों की आवश्यकता होती है
(क) कम्प्यूटर तथा अन्य बड़ी मशीनों की।
(ख) सहायता के लिये अनेक प्रशिक्षित व्यक्तियों की।
(ग) सूक्ष्मदर्शी, दूरदर्शी तथा परखनलियों की।
(घ) विभिन्न प्रकार के विशेष उपकरणों की।

12. वैज्ञानिक बहुधा बहुत कठिन परिश्रमी और अपने कार्य के लिये पूर्ण समर्पित कहे जाते हैं, यह बात सत्य है
(क) किसी भी कार्य में सफलता प्राप्त लोगों के विषय में।
(ख) वैज्ञानिकों के विषय में न कि दूसरे कार्यों में लगे लोगों के विषय में।
(ग) महत्वपूर्ण कार्यों में लगे अधिकतर लोगों के विषय में न कि वैज्ञानिकों के विषय में।
(घ) कुछ उच्च कोटि के वैज्ञानिकों के प्रायोगिक कार्यों के विषय में।

13. जब बहुत से खोजे गये नये तथ्य किसी वैज्ञानिक सिद्धान्त की पुष्टि नहीं करते हैं तो वैज्ञानिक प्राय: स्वयं से प्रश्न करते हैं
(क) क्या हमें इस सिद्धान्त को रद्द कर देना होगा क्योंकि नये तथ्य इसकी पुष्टि नहीं करते हैं।
(ख) क्या हम इन तथ्यों में कुछ परिवर्तन कर सकते है ताकि ये वैज्ञानिक सिद्धान्त के अनुकूल हो जाएँ।

(ग) क्या हम इस सिद्धान्त को इसी प्रकार छोड़ दें क्योंकि नये तथ्य इसके अनुकूल नहीं हैं।

(घ) क्या हम इस सिद्धान्त में थोड़ा परिवर्तन कर सकते हैं ताकि नये तथ्य इसके अनुकूल हो जायें।

14. वैज्ञानिक अधिकतर महत्वपूर्ण खोज

(क) बहुत से प्रश्नों पर विचार करके करते हैं।

(ख) प्रयोगों का अध्ययन करके करते हैं।

(ग) विचारों की छान-बीन करके करते हैं।

(घ) विभिन्न प्रेक्षणों (Observations) के पश्चात् करते हैं।

15. जनता के समक्ष एक नये सिद्धान्त की घोषणा से पहले इस बात की पर्याप्त सम्भावना है कि वैज्ञानिक अपने इन विचारों पर

(क) अपने ही विशेष क्षेत्र के अन्य वैज्ञानिकों से चर्चा करना चाहेंगे।

(ख) विभिन्न विज्ञान-पत्रिकाओं के सम्पादकों से चर्चा करना चाहेंगे।

(ग) वैज्ञानिक सिद्धान्तों के विशेषज्ञों के ग्रुप के साथ चर्चा करना चाहेंगे।

(घ) अपने सिद्धान्त में रुचि दिखाने वाले सरकारी नेताओं के साथ चर्चा करना चाहेंगे।

16. निम्न में से कौन सा वाक्य विज्ञान के विषय में सर्वोत्तम है?

(क) आधुनिक विज्ञान इतना विकसित हो चुका है कि वह भूतकाल के आविष्कारों का उपयोग नहीं कर सकता।

(ख) आधुनिक विज्ञान आधुनिक उत्पादन विकसित करता है।

(ग) आधुनिक विज्ञान उपयोगी आविष्कारों पर निर्भर है।

(घ) आधुनिक विज्ञान भूतकालीन विज्ञान पर आधारित है।

17. अमेरिकी वैज्ञानिक अपने अनुसंधान कार्यों के लिए अत्यधिक धन की मांग करते हैं। इसका मुख्य कारण है कि

(क) जितना अधिक धन दूसरे कार्यों पर खर्च होता है उतना ही विज्ञान के कार्यों पर खर्च होना चाहिये।

(ख) बहुत सी वैज्ञानिक परियोजनाओं के लिए लम्बे समय और अधिक उपकरणों की आवश्यकता होती है।

(ग) 1945 के बाद से वैज्ञानिक सामग्री एवं उपकरणों के मूल्यों में वृद्धि होती रही है।

(घ) नई वैज्ञानिक प्रयोगशालाएँ बड़ी होती है तथा इनके निर्माण में अधिक खर्च आता है।

18. जब एक वैज्ञानिक किसी वैज्ञानिक आविष्कार की रिपोर्ट पढ़ता है तो सम्भवत:
    (क) वह उस पर पूरा विश्वास नहीं करेगा जब तक कि वह स्वयं उसकी जांच न कर ले।
    (ख) इस पर विश्वास कर लेगा क्योंकि रिपोर्ट वैज्ञानिकों के कार्यों का विवरण है।
    (ग) वह इस पर पूर्णत: विश्वास नहीं करेगा जब तक कि इस विषय में उसे और अधिक जानकारी नहीं मिल जाती।
    (घ) वह रिपोर्ट पर विश्वास कर लेगा क्योंकि वैज्ञानिक रिपोर्ट में कोई त्रुटि नहीं होती।

19. वैज्ञानिकों को अपने कार्यों में अधिक मापन करना होता है निम्न में से किस अवस्था में वैज्ञानिक मापन कार्य अधिक करना चाहेगा।
    (क) एक गैलन पेट्रोल में एक मीटर कार कितनी दूरी तय करती है।
    (ख) सर्दियों में पक्षी दक्षिण की ओर कितनी दूर जाते हैं।
    (ग) सुरक्षित रहने तक एक पुल कितना वजन सह सकता है।
    (घ) कोई टूथ पेस्ट/दंतमंजन कितने कीटाणुओं का नाश कर सकता है।

20. सीता अपने वैज्ञानिक प्रयोग के बारे में सब कुछ विस्तारपूर्वक लिखती है। यदि सीता एक वैज्ञानिक बन जाये तो वह प्रशिक्षण उसका सहायक होगा
    (क) अपने प्रयोग को धैर्यपूर्वक करने में।
    (ख) अपने प्रयोग के विषय में अधिक अच्छी रिपोर्ट देने में।
    (ग) अपने प्रयोग के आधार पर सिद्धान्त सोच निकालने में।
    (घ) नये प्रयोग करने की राह सुझाने में।

21. निम्न में से कौनसा वैज्ञानिकों के मुख्य कार्यकलापों (Activity) का वर्णन है?
    (क) वैज्ञानिक कुशल सेवा और सलाह प्रदान करते हैं।
    (ख) वैज्ञानिक हमें अधिक स्वस्थ रहने की राह दिखाते हैं।
    (ग) वैज्ञानिक प्रकृति के बारे में समझने का विकास करते हैं।
    (घ) वैज्ञानिक अच्छे जीवनयापन, जीविका के लिये अच्छी वस्तुओं का उत्पादन करते हैं।

22. कमला सदा विद्यालय में अच्छा करती है, हवाई जहाज़ का मॉडल बनाना पसन्द करती है तथा अपने सहपाठियों के साथ हँसी मजाक करती है। संजय को गणित के अच्छे नंबर (अंक) आते हैं, वह किताबें पढ़ना पसन्द करता है और फुटबॉल खेलता है। सरला एक धीर चिन्तनशील और चतुर बालिका है तथा नाचना पसन्द करती है। आपके विचार में किसके वैज्ञानिक बन सकने की सम्भावना है?

(क) कमला के।
(ख) संजय के।
(ग) सरला के।
(घ) उपयुक्त में किसी के भी।

23. सूक्ष्मदर्शी ने वैज्ञानिकों के लिए छोटी चीज़ों का देखना सम्भव बनाने के साथ ही क्या और सम्भव बना दिया?
(क) बहुत सी नई समस्याओं की खोज करना।
(ख) जीवन अर्थ को समझना।
(ग) गतिशील सूक्ष्म कणों का निरीक्षण।
(घ) बीमारी पैदा करने वाले कीटाणुओं को देखना।

24. ,d oSKkfud fu;e (Scientific Law) व्याख्या करता है
(क) उन नियमों की जिनका वैज्ञानिक अवश्य पालन करते हैं।
(ख) उन नियमों को जो प्रकृति की घटनाओं से संबंध स्थापित करते हैं।
(ग) अच्छे प्रयोग करने से सम्बन्धित नियमों की।
(घ) घटनाओं के कारण के विषय में अच्छे अनुमान की।

25. निम्नलिखित में से कौनसा वाक्य विज्ञान की सबसे अच्छी व्याख्या करता है।
(क) विज्ञान प्रयोग करने का नाम है।
(ख) विज्ञान एक विचार एवं योजना है।
(ग) विज्ञान चिन्तन एवं करनी का नाम है।
(घ) विज्ञान प्रेक्षण एवं मापन है।

26. एक वैज्ञानिक अपने कार्य के प्रति उदार-मनोवृत्ति (Open Mind) का होता है। यदि वह
(क) दूसरे वैज्ञानिकों के विचारों से भी सहमत होता है।
(ख) अपने विचारों के विरुद्ध विचारों पर ध्यान देता है।
(ग) प्रयोगों के विषय में बहुत सारे नये विचार सोचता है।
(घ) अपने अधिकतर विचारों पर दूसरे वैज्ञानिकों के साथ चर्चा करता है।

27. वैज्ञानिकों द्वारा पौधों का अध्ययन करने का मुख्य कारण
(क) खाद्य पदार्थ की उपज बढ़ाने में किसानों की सहायता करना है।
(ख) नयी दवायें बनाने के तरीकों की खोज करना है।
(ग) पौधों के जीवन तथा वृद्धि को समझना है।
(घ) उन स्थानों का पता लगाना है जहाँ वे अच्छी तरह उग सकते हैं।

28. निम्नलिखित में से विज्ञान की मुख्य आवश्यकता क्या है?
    (क) नये विचार के लोग।
    (ख) अधिक धन और साधन।
    (ग) पूर्णतया प्रशिक्षित कर्मचारी।
    (घ) अच्छी कार्य परिस्थितियाँ।

29. जब हम कहते हैं कि एक वैज्ञानिक ने किसी प्रयोग के विषय में परिकल्पना (Hypothesis) की है तो हमारे कहने का अर्थ होता है कि
    (क) उसने यह संकेत दिया है कि क्या-क्या मापन करना होगा।
    (ख) उसने प्रयोग के लिए आवश्यक उपकरणों का डिज़ाइन अभिकल्प बनाया है।
    (ग) उसका प्रयोग किस प्रकार हो पायेगा, इसकी रिपोर्ट दी है।
    (घ) उसने आगे होने वाली घटना का सावधानी पूर्वक अनुमान लगाया है।

30. निम्न में से कौन-सी सूची सबसे उपयुक्त है जिसका वैज्ञानिक अध्ययन करते हैं?
    (क) परमाणु, अणु और तारे।
    (ख) पदार्थ, उर्जा और सजीव।
    (ग) बनस्पति, पौधे, पशु-पक्षी और रोग।
    (घ) रॉकेट, उपग्रह और अंतरिक्ष यात्रा।

31. जब कोई वैज्ञानिक एक नये वैज्ञानिक सिद्धान्त (Scientific Theory) का प्रतिपादन करता है तो हमें पता है कि उसने
    (क) प्रकृति के नियमों में से एक को अभिव्यक्त किया है।
    (ख) मानव जाति को सच्चाई की ओर अग्रसर करने में सहायता की है।
    (ग) प्रयोगों को नये ढंग से करने के तरीके की खोज की है।
    (घ) नये विचारों एवं समझ को विकसित किया है।

32. क्या हमें एक नये प्रकार की मोटर गाड़ी तैयार करने की योजना बनाने वाले व्यक्ति को वैज्ञानिक कहना चाहिए?
    (क) हाँ, क्योंकि वह अपनी योजना में वैज्ञानिक विधि का प्रयोग करता है।
    (ख) नहीं, क्योंकि वह एक उपयोगी मशीन बनाने की योजना बनाता है।
    (ग) हाँ, क्योंकि वह अपनी योजना की जाँच के लिए प्रयोग करता है।
    (घ) नहीं, क्योंकि वह कुछ नये विचारों को परखने के लिए योजना बनाता है।

33. एक वैज्ञानिक सिद्धान्त को
    (क) वैज्ञानिक प्रश्नों का अन्तिम निर्णयात्मक उत्तर देना चाहिए।

(ख) अधिक उपयोगी वस्तुएँ बनाने का मार्गदर्शन करना चाहिए।
(ग) विभिन्न घटनाओं का समन्वय तथा उनकी व्याख्या करनी चाहिए।
(घ) प्रयोगों को करने के लिए अच्छे नियमों, तरीकों का सुझाव देना चाहिए।

34. जब कोई वैज्ञानिक एक नयी खोज करता है तो प्रायः वह उसकी रिपोर्ट तैयार करता है। वह ऐसा करता है क्योंकि
(क) उसे आशा होती है कि दूसरे वैज्ञानिक उसके कार्य को पूरा करने में उसकी सहायता करेंगे।
(ख) वह चाहता है कि दूसरे वैज्ञानिक उसके काम के बारे में जाने तथा उसकी जाँच करें।
(ग) उसे आशा होती है कि उसकी कार्य-समाप्ति रिपोर्ट से उसके दूसरे साथियों को सहायता मिलेगी।
(घ) वह जानना चाहता है कि क्या दूसरों ने भी इसी कार्य को किया है तथा इसकी रिपोर्ट दी हैं।

35. विज्ञान में प्रायोगः
(क) मनुष्य की समस्याओं के समाधान के लिए किये जाते हैं।
(ख) प्रकृति के विषय में सच्चाई का पता लगाने के लिए किये जाते हैं।
(ग) वैज्ञानिकों के विचारों को परखने के लिए किये जाते हैं।
(घ) ब्रह्माण्ड को क्रमबद्ध सिद्ध करने के लिए किये जाते हैं।

36. अधिकतर वैज्ञानिक चुस्त एवं चतुर (Smart) होते हैं। वे साधारण लोगों की अपेक्षा अधिक सरलता से सीख लेते हैं तथा अपने मस्तिष्क के उपयोग से अधिक कठिन कार्य कर सकते हैं, क्या यह कथन सत्य है।
(क) हाँ, परन्तु डॉक्टर तथा वकीलों से अधिक चुस्त एवं चतुर नहीं होते।
(ख) हाँ, परन्तु केवल इसलिए कि वे वैज्ञानिक-कुशलता के साथ जन्म लेते हैं।
(ग) हाँ, परन्तु केवल इसलिए कि वैज्ञानिकों को विशेष निर्देश दिये जाते है।
(घ) नहीं, वैज्ञानिक अन्य लोगों के बराबर ही चुस्त एवं चतुर होते हैं उनसे अधिक नहीं।

37. निम्नलिखित वाक्यों में से कौनसा वाक्य प्रौद्योगिकी का सर्वोचित रूप सबसे उपयुक्त तरह से परिभाषित करता है।
(क) वह ज्ञान जिससे हमें विभिन्न मौसमों के विषय में जानकारी मिलती है।
(ख) वह ज्ञान जो विज्ञान से मिलता-जुलता है।
(ग) वह ज्ञान जो नये तरीकों के लिए, औजारों और मशीनों को बनाने के लिए इस्तेमाल होता है।

(घ) काम करने का ऐसा अलग तरीका जिससे दूसरे प्रभावित हों।

38. विज्ञान और प्रौद्योगिकी एक-दूसरे से संबंधित हैं क्योंकि
   (क) विज्ञान की जानकारी से ही प्रौद्योगिकी का इस्तेमाल होता है और प्रौद्योगिकी को बढ़ावा देने से विज्ञान का ज्ञान भी बढ़ता है।
   (ख) विज्ञान सभी तकनीकी उपलब्धियों की जननी है।
   (ग) विज्ञान और तकनीकी के पारस्परिक नज़दीकी संबंधों के कारण दोनों को अलग कर पाना कठिन है।
   (घ) विज्ञान हमें बताता है कि जीवन क्या है तथा तकनीक हमें जीने के विभिन्न तरीकों के बारे में बताती है।

39. भारत में जीवन स्तर को ऊंचा उठाने के लिए क्या ज्यादा जरूरी है–तकनीकी की प्रगति में किया गया खर्च या विज्ञान के क्षेत्र में लगाया गया धन
   (क) दोनों ही ज़रूरी हैं क्योंकि दोनों में कोई भी अन्तर नहीं है।
   (ख) दोनों में से कोई भी ज़रूरी नहीं है क्योंकि जीवन स्तर इन दोनों से नहीं सुधर पाएगा।
   (ग) दोनों को बराबर महत्व देना चाहिए क्योंकि दोनों एक-दूसरे के पूरक हैं और समाज में अपने-अपने तरीके से फायदेमन्द हैं।
   (घ) हमें चिकित्सा और पर्यावरण से संबंधित वैज्ञानिक खर्च करने चाहिएं क्योंकि यह नई मशीनें और औज़ार आदि बनाने पर होने वाले प्रौद्योगिकीय खर्च से ज्यादा जरूरी है।

40. प्रौद्योगिकी विज्ञान पर निर्भर करती है क्योंकि
   (क) प्रौद्योगिकी का ज्ञान विज्ञान से ही उत्पन्न होता है।
   (ख) किसी मशीन का पुर्जा बनाने या नल ठीक करने के लिए हमें किसी वैज्ञानिक खोज की ज़रूरत होती है।
   (ग) विज्ञान और प्रौद्योगिकी दोनों एक ही ज्ञान पर आधारित है।
   (घ) प्रौद्योगिकी वैज्ञानिक खोजों के सहारे के बिना प्रगति करती है।

41. वैज्ञानिकों को अपनी खोजों के हानिकारक और लाभदायक प्रभावों को ध्यान में रखना चाहिए क्योंकि
   (क) वैज्ञानिकों की खोजों का दूरगामी एवं गहरा प्रभाव समाज पर पड़ता है।
   (ख) विज्ञान ही वर्तमान समाज का संचालन करता है।
   (ग) केवल वैज्ञानिक ही अपनी खोजों के लाभदायक और हानिकारक प्रभावों का सही निर्णय ले सकते हैं।
   (घ) वैज्ञानिकों को अपनी खोजों के सभी प्रभावों की जानकारी नहीं होती है।

42. यह वैज्ञानिकों की जिम्मेदारी है कि वे अपनी खोजों के बारे में इस तरह से बतायें कि आम जनता उन्हें समझ सके
    (क) क्योंकि वैज्ञानिक खोजें आम आदमी के लिए बहुत मुश्किल और उलझी हुई होती हैं।
    (ख) ताकि जनता को पता चल सके कि जनता का पैसा विज्ञान पर किस प्रकार खर्च होता है।
    (ग) ताकि जनता वैज्ञानिक खोजों की जानकारी से अपना जीवन स्तर सुधार सके और अपने भविष्य पर पड़ने वाले प्रभावों का पता लगा सके।
43. परमाणु निरस्तीकरण का निर्णय केवल वैज्ञानिकों एवं इंजीनियरों को ही नहीं लेना चाहिए क्योंकि
    (क) उनका प्रशिक्षण और जानकारी उन्हें इस विषय की ज्यादा समझ नहीं देती है।
    (ख) अपनी जानकारी के कारण वे निर्णय लेने में ज्यादा सक्षम नहीं होते, जबकि सरकार और प्राइवेट कम्पनियाँ इस पर बेहतर निर्णय ले सकती है।
    (ग) हमारे समाज को प्रभावित करने वाले निर्णयों में वैज्ञानिको, इंजीनियरों, विशेषज्ञों एवं जनता सबका बराबर का योगदान होना चाहिए।
    (घ) यह एक राजनीतिक विषय है और केवल सरकार को इसका निर्णय लेना चाहिए।
44. विज्ञान और प्रौद्योगिकी से अनेक समस्याओं को सुलझाने में मदद मिलती है जैसे गरीबी, अराजकता, बेकारी, जनसंख्या नियंत्रण व परमाणु युद्ध आदि।
    (क) हाँ, क्योंकि विज्ञान के सभी नये विचार और प्रौद्योगिकी के सभी नये तरीके, समस्याओं को सुलझाने में मदद करते हैं।
    (ख) विज्ञान एवं प्रौद्योगिकी से केवल कुछ ही सामाजिक समस्याएं सुलझायी जा सकती हैं, सभी नहीं।
    (ग) विज्ञान एवं प्रौद्योगिकी कुछ सामाजिक समस्याऐं सुलझाती है, तो वही अनेक समस्याओं को जन्म भी देती है।
    (घ) समस्या विज्ञान एवं प्रौद्योगिकी से मदद लेने की नहीं बल्कि लोगों द्वारा उसे बुद्धिमतापूर्वक इस्तेमाल करने की है।

प्रश्न संख्या 45 से प्रश्न संख्या 50 तक प्रत्येक प्रश्न में चार विकल्प दिए गए हैं जिनमें से तीन विकल्प सही हैं और एक विकल्प गलत है। गलत विकल्प का चयन करें और उस पर सही का निशान लगाएँ।

45. वैज्ञानिक प्रतिदिन की समस्याओं को बेहतर ढंग से सुलझा सकते हैं। जैसे किसी

कार को गड्ढे से बाहर निकालना, खाना बनाना या पालतू पशुओं की देखभाल करना।

(क) उन्हें वैज्ञानिक ढंग से प्रशिक्षण मिला होता है और उनका लोजिकल तर्कसंगत दिमाग होता है।

(ख) उनकी औपचारिक शिक्षा उनके विशिष्ट ज्ञान को विकसित करती है।

(ग) उनका अनुभव और सामान्य ज्ञान कामन सन्स रोजमर्रा की समस्याएँ सुलझाने में मदद करता है।

(घ) वे अधिक बुद्धिमान और दूसरों से भिन्न होते हैं।

46. तकनीकी के ज्यादा इस्तेमाल से किसी भी देश के लोगों का जीवन स्तर सुधर जायेगा

(क) हाँ, क्योंकि प्रौद्योगिकी हमेशा जीवन स्तर सुधारती है और अब भी ऐसा कोई कारण नहीं है जो उसे ऐसा करने से रोके

(ख) हाँ, क्योंकि हम जितना ज्यादा जानेंगे उतनी ही अच्छी तरह हम अपनी देखभाल कर पाएँगे और अपनी समस्याएँ सुलझा सकेंगे।

(ग) हाँ, क्योंकि प्रौद्योगिकी से नयी नौकरियाँ और तरक्की के अवसर मिलते हैं।

(घ) हाँ भी और नहीं भी, क्योंकि प्रौद्योगिकी के अधिक इस्तेमाल से जहाँ जीवन ज्यादा आरामदायक और स्वस्थ बनता है वहीं इससे कहीं ज्यादा प्रदूषण, बेकारी और अन्य समस्याएँ भी उठ खड़ी होती हैं।

47. सैन्यशक्ति विज्ञान और प्रौद्योगिकी पर काफी हद तक निर्भर करती है

(क) क्योंकि विज्ञान और प्रौद्योगिकी का जितना ज्यादा विकास होगा उतना ही ज्यादा आधुनिक, सटीक और विनाशकारी हथियार होगा।

(ख) यह इस बात पर निर्भर करता है कि सैन्य अफसरों में कितनी योग्यता है और वे सरकार पर कितना दबाव डाल सकते हैं।

(ग) किसी सेना की शक्ति केवल विज्ञान और प्रौद्योगिकी से बने शक्तिशाली हथियारों पर ही नहीं बल्कि सेना की संख्या पर भी निर्भर करती है।

(घ) सैन्यशक्ति विज्ञान और तकनीकी पर नहीं बल्कि सरकार पर निर्भर करती है।

48. किसी भी गणतंत्र में वैज्ञानिकों द्वारा विभिन्न समस्याओं का चुनाव इस बात पर निर्भर करता है कि

(क) उस देश की सामाजिक जरूरतें क्या हैं?

(ख) उस देश को किस प्रकार की प्रौद्योगिकी उपलब्ध हैं।

(ग) उस देश की वैज्ञानिकों की शिक्षा कैसी है?

(घ) उस देश के पड़ोसी देशों में किस प्रकार की खोजें की गई हैं।

49. यदि कोई नई प्रौद्योगिकी विकसित होती है तो उसे अपनाने के लिए जरूरी है कि
   (क) प्रौद्योगिकी पर्यावरण के अनुकूल हो।
   (ख) प्रौद्योगिकी लागत की दृष्टि से किफायती हो।
   (ग) प्रौद्योगिकी कार्यक्षम हो।
   (घ) प्रौद्योगिकी सरकारी आमदनी को बढ़ाती हो।

50. प्रौद्योगिकी विकास नागरिकों द्वारा नियंत्रित हो सकता है।
   (क) हाँ, तकनीकी विकास सरकार द्वारा चलाया जाता है और क्योंकि नागरिक ही सरकार चुनते हैं, इस प्रकार वे तकनीकी विकास पर भी नियंत्रण रखते हैं।
   (ख) हाँ, तभी जब नये प्रौद्योगिकी विकास का हानिकारक प्रभाव जनता पर पड़ता है।
   (ग) नहीं, क्योंकि तकनीकी विकास इतनी तेजी से आगे बढ़ता है कि एक आम आदमी इस विकास के बारे में अनजान होता है।
   (घ) नहीं, क्योंकि प्रौद्योगिकी विकास का प्रभाव जनता की समझ से बाहर है।

APPENDIX D

Pre-test and Post-test

पूर्वज्ञान परीक्षा

कुल 30 प्रश्न हैं 1 प्रत्येक प्रश्न में चार विकल्प हैं, प्रत्येक प्रश्न में सबसे उचित विकल्प पर सही का निशान लगायें।

1. जब हम किसी जगह की जनसंख्या की बात करते हैं तो हमारा अभिप्राय वहाँ रहने वाले
   (क) पुरुषों की संख्या से होता है।
   (ख) स्त्रियों की संख्या से होता है।
   (ग) बच्चों की संख्या से होता है।
   (घ) उपयुक्त तीनों की संख्या से होता है।

2. मान लो महेन्द्रगढ़ की जनसंख्या एक करोड़ साठ लाख है, खोल की तीस लाख है और रेवाड़ी की डेढ़ करोड़ पैंतीस लाख है। तो किस जगह की जनसंख्या सबसे अधिक है।
   (क) मेहन्द्रगढ़।
   (ख) खोल।
   (ग) रेवाड़ी।
   (घ) मेहन्द्रगढ़ और खोल।

3. संक्रामक रोग किस प्रकार के क्षेत्रों में अधिक तेजी से फैलता है।
   (क) घनी आबादी वाले क्षेत्रों में।
   (ख) कम आबादी वाले क्षेत्रों में।
   (ग) दोनों में से किसी में नहीं।
   (घ) दोनों क्षेत्रों में बराबर।

4. कैसा परिवार ज्यादा सुखी रहता है।
   (क) बड़ा परिवार।
   (ख) छोटा परिवार।
   (ग) बहुत सारे रिश्तेदार वाला परिवार।
   (घ) कम रिश्तेदार वाला परिवार।

5. एक ही विद्यालय की यदि कक्षा 7(अ) में 80 बच्चे हैं और कक्षा 7(ब) में 50 बच्चे हैं तो किस कक्षा के बच्चे सुविधापूर्वक पढ़ाई कर पाएंगे।
   (क) 7(अ) के।
   (ख) 7(ब) के।
   (ग) दोनों 7(अ) और 7(ब) के।
   (घ) दोनों कक्षाओं में से किसी के नहीं।

6. यदि आप के गांव की जनसंख्या डेढ़ हजार है और आपके मोहल्ले की साठ तो आपके मोहल्ले की जनसंख्या आप के गांव की जनसंख्या की तुलना में कितनी है।
   (क) 4%
   (ख) 3.4%
   (ग) 5.7%
   (घ) 2.8%

7. यदि भारत की जनसंख्या 68.5 करोड़ है और हरियाणा की 1.29 करोड़ तो हरियाणा की जनसंख्या भारत की जनसंख्या की तुलना में कितनी प्रतिशत है।
   (क) 2.43%
   (ख) 3.45%
   (ग) 1.93%
   (घ) 2.8%

8. हस्पताल में मरीजों की संख्या बढ़ने से
   (क) सरकार को ज्यादा पैसा खर्च करना पड़ेगा।
   (ख) डॉक्टरों पर काम का बोझ बढ़ जाएगा।
   (ग) डॉक्टर मरीज़ों की देखभाल अच्छी तरह से नहीं कर सकेंगे।
   (घ) उपरोक्त सभी।

9. आपके गांव में यदि दो किसानों के पास बराबर के खेत हैं और इनमें से एक किसान के दो बेटे हों और दूसरे किसान के पांच बेटे तो किसके बेटों को ज्यादा खेत मिलेंगे।
   (क) पहले किसान के बेटों को।
   (ख) दूसरे किसान के बेटों को।
   (ग) दोनों किसानों के बेटों को बराबर खेत मिलेंगे।
   (घ) बेटों की संख्या ज्यादा होने से कोई फर्क नहीं पड़ेगा।

10. हमारे देश की जनसंख्या दुगनी हो जाने पर

(क) हमारे देश में मानव संसाधन अधिक हो जाएँगे।
(ख) सस्ते दरों पर मजदूर मिलने लगेंगे।
(ग) शहरी इलाकों में समस्याएँ बढ़ जाएँगी।
(घ) उपरोक्त सभी।

11. शाकाहारी जीवों की संख्या बढ़ने से
(क) पेड़-पौधों की कमी होगी।
(ख) पर्यावरण की सुरक्षा होगी।
(ग) मांसाहारी जीव बढ़ जाएँगे।
(घ) मनुष्यों को शिकार करने में सुविधा होगी।

12. किसी जगह सूखा पड़ने से
(क) लोगों को कम काम करना पड़ेगा।
(ख) लोग भूख से मरने लगेंगे।
(ग) लोग ज्यादा खेती करेंगे।
(घ) उपरोक्त में से कोई नहीं।

13. जहाँ रोजगार की सुविधा न हो, वहीं लोग
(क) अधिक रहेंगे।
(ख) कम रहेंगे।
(ग) खुश रहेंगे।
(घ) ज्यादा काम करेंगे।

14. मनुष्यों की संख्या बढ़ने से
(क) जीव-जन्तुओं की संख्या भी बढ़ती हैं।
(ख) ज्यादा खेती होती हैं।
(ग) पेड़-पौधे अधिक होते हैं।
(घ) प्रकृति का संतुलन बिगड़ जाता है।

15. नीचे दिए गए पिक्टोग्राफ (च), (छ) और (ज) में से कौन सा पिक्टोग्राफ ज्यादा जनसंख्या दर्शा रहा है

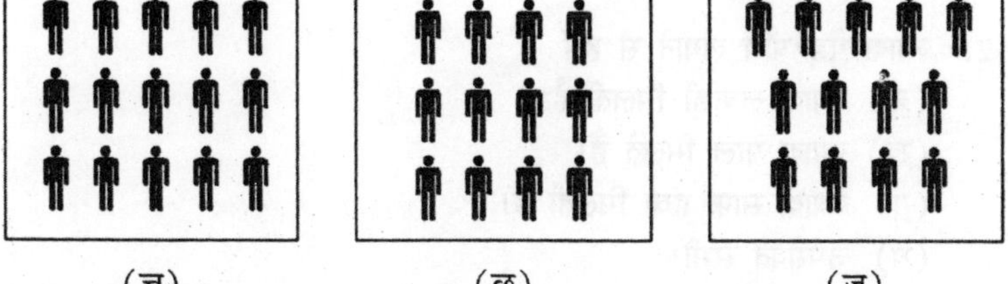

(च) (छ) (ज)

(क) पिक्टोग्राफ च
(ख) पिक्टोग्राफ छ
(ग) पिक्टोग्राफ ज
(घ) तीनों पिक्टोग्राफ बराबर की जनसंख्या दर्शा रहे हैं।

16. निम्न में से कौन-सा वाहन वायु में धुंआ फैलाता है?
(क) बैलगाड़ी (ख) तांगा
(ग) ट्रक (घ) साईकल

17. तालाब, नल और कुएं में से किस का पानी अधिक साफ होगा
(क) नल का (ख) तालाब का
(ग) कुएँ का (घ) तीनों का

18. कल-कारखानों से निकलने वाला धुआं
(क) स्वास्थ्य के लिए लाभदायक है।
(ख) स्वास्थ्य के लिए हानिकारक है।
(ग) स्वास्थ्य पर कोई प्रभाव नहीं डालता है।
(घ) पेड़ पौधों के लिए लाभदायक होता है।

19. धूम्रपान करने से कौन-कौन प्रभावित होते हैं
(क) धूम्रपान करने वाला व्यक्ति प्रभावित होता है।
(ख) आस-पास के सभी व्यक्ति प्रभावित होते हैं।
(ग) किसी पर भी बुरा प्रभाव नहीं पड़ता।
(घ) धूम्रपान करने वाला व्यक्ति और आसपास के सभी व्यक्ति प्रभावित होते हैं।

20. तेज़ आवाज़ में टेलीविज़न एवं रेडियो बजाने से
(क) सबका मनोरंजन होता है।
(ख) स्मरण शक्ति बढ़ती है।
(ग) कान खराब हो जाते हैं।
(घ) अच्छा सुनाई देता है।

21. ज्यादा पेड़-पौधे लगाने से हमें
(क) ज्यादा लकड़ी मिलती है।
(ख) ज्यादा फल मिलते हैं।
(ग) ज्यादा साफ हवा मिलती है।
(घ) उपरोक्त सभी।

22. स्वच्छ पर्यावरण किनके लिए लाभदायक होता है
    (क) मनुष्य के लिए। (ख) पशुओं के लिए।
    (ग) पौधों के लिए। (घ) सभी के लिए।

23. पर्यावरण किससे बनता है
    (क) मिट्‌टी, पानी, हवा, जीव-जन्तु और पेड़-पौधे से।
    (ख) केवल हवा से।
    (ग) केवल जीव-जन्तुओं से।
    (घ) केवल पेड़-पौधों से।

24. यदि हम पेड़-पौधों को काट देंगे तो हवा में कौन-सी गैस की कमी हो जाएगी
    (क) कार्बन डाईऑक्साईड।
    (ख) ऑक्सीजन।
    (ग) कार्बन मोनोऑक्साइड।
    (घ) नाइट्रोजन।

25. लकड़ी से जलने वाला चूल्हा और गोबर गैस से जलने वाले चूल्हें में से कौन-सा चूल्हा ज्यादा धुआं फैलाता है
    (क) लकड़ी से जलने वाला चूल्हा।
    (ख) गोबर गैस से जलने वाला चूल्हा।
    (ग) दोनों में से कोई भी नहीं।
    (घ) दोनों ही धुआं फैलाते हैं।

26. प्रकाश रांश्लेपक के लिए पेड़-पौधों को कौन-सी गैस की आवश्यकता होती है
    (क) आक्सीजन (ख) कार्बन डाइऑक्साइड
    (ग) नाइट्रोजन (घ) कार्बन मोनोऑक्साइड

27. मरे हुए जीव-जन्तुओं एवं कूड़े-करकट को नदी में बहा देने से
    (क) वातावरण की सफाई होती है।
    (ख) कूड़े-करकट से छुटकारा मिलता है।
    (ग) पानी गंदा हो जाता है।
    (घ) खर्चा कम होता है।

28. रोगी को खांसने, छींकने से किस रोग के रोगाणु हवा में फैल जाते हैं
    (क) तपेदिक। (ख) टिटनस।
    (ग) मलेरिया। (घ) पायरिया।

29. दूषित जल से फैलता है
    (क) हैजा। (ख) चेचक।
    (ग) पीलिया। (घ) क और ग।

30. पेड़-पौधों पर कीटनाशकों का प्रयोग किसलिए करते हैं
    (क) खुशबू फैलाने के लिए।
    (ख) पेड़-पौधों को कीड़ों से बचाने के लिए।
    (ग) सिंचाई के लिए।
    (घ) सिर्फ मच्छर मारने के लिए।

ज्ञान परीक्षा

कुल 30 प्रश्न हैं। प्रत्येक प्रश्न में चार विकल्प हैं, प्रत्येक प्रश्न में सबसे उचित विकल्प पर सही का निशान लगायें।

1. पर्यावरण का संतुलन बिगड़ने को ही .... कहते हैं
    (क) प्रदूषण। (ख) जनसंख्या वृद्धि।
    (ग) भूकम्प। (घ) बाढ़।

2. जनसंख्या बढ़ने से पर्यावरण में कौन सी गैस की अधिकता हो रही है
    (क) कार्बन डाईऑक्साइड। (ख) ऑक्सीजन।
    (ग) नाइट्रोजन। (घ) हाइड्रोजन।

3. जनसंख्या वृद्धि के कारण वायुमंडल में किस गैस की कमी होती जा रही है।
    (क) ऑक्सीजन। (ख) कार्बन डाईऑक्साइड।
    (ग) हाइड्रोजन। (घ) नाइट्रोजन।

4. वायु प्रदूषण कम किया जा सकता है
    (क) कारखानों की चिमनियों को ऊँचा करके।
    (ख) चिमनियों को हटाकर।
    (ग) चिमनियों को कम ऊँचाई पर लगाकर।
    (घ) चिमनियों को पतला बनाकर।

5. जिस स्थान पर कल-कारखाने नहीं होते वहां प्रदूषण .... होता है
    (क) कम। (ख) अधिक।
    (ग) बिल्कुल नहीं। (घ) रोजाना।

6. ध्वनि प्रदूषण से .... रोग होता है
   (क) हैजा।	(ख) बहरापन।
   (ग) टी.बी.।	(घ) तपेदिक।

7. धूम्रपान एक कारण है
   (क) वायु प्रदूषण का।	(ख) जल प्रदूषण का।
   (ग) ध्वनि प्रदूषण का।	(घ) मृदा प्रदूषण का।

8. मानव और प्रकृति में तालमेल बिगड़ गया है क्योंकि
   (क) वैज्ञानिक ज्ञान का अनुपयुक्त उपयोग हुआ है।
   (ख) प्राकृतिक संसाधनों का अंधाधुन्ध शोषण हुआ है।
   (ग) मानव लालची है।
   (घ) उपरोक्त सभी।

9. पौधों व जन्तुओं के अवशिष्ट पदार्थों से उपयोगी पदार्थ बनाए जा सकते हैं, यदि उनको
   (क) कम्पोस्ट में बदल दिया जाए।
   (ख) इधर-उधर फैंक दिया जाए।
   (ग) तुरन्त जला दिया जाए।
   (घ) नदी में बहा दिया जाए।

10. बेकार पानी के निबटारे का सबसे उत्तम उपाय कौनसा है
   (क) खुले गढ्ढे में डालना।
   (ख) खुली नालियों में बहाना।
   (ग) आंगन वाटिका के पौधों में डालना।
   (घ) पशुओं को पिलाना।

11. पौधों तथा जन्तुओं का संरक्षण करना चाहिए क्योंकि वे
   (क) देखने में सुन्दर लगते हैं।
   (ख) पास-पड़ोस की सुन्दरता बढ़ाते हैं।
   (ग) प्रकृति में संतुलन बनाए रखते हैं।
   (घ) हमारे भोजन के लिए लाभकारी हैं।

12. मृदा अपरदन को नियंत्रित किया जा सकता है
   (क) केवल बांध बनाकर।
   (ख) केवल वृक्ष लगाकर।
   (ग) केवल वेदिकाएँ बनाकर।

(घ) ऊपर के सभी कार्य करके।

13. ताप बिजलीघर की तुलना में जल बिजलीघर को प्राथमिकता दी जानी चाहिए क्योंकि
    (क) इनसे प्रदूषण नहीं होता।
    (ख) ये कम खर्च पर ऊर्जा उत्पन्न करते हैं।
    (ग) बहुमूल्य प्राकृतिक संसाधनों की बचत में सहायता करते हैं।
    (घ) उपरोक्त सभी।

14. यदि अपशिष्ट पदार्थ जल-स्रोत को संदूषित कर दें तो कौन-सा रोग फैल सकता है
    (क) अरक्तता। (ख) हैजा।
    (ग) रतौंधी। (घ) तपेदिक।

15. ध्वनि प्रदूषण फैलता है
    (क) तेज शोर से। (ख) हल्के शोर से।
    (ग) पेड़ काटने से। (घ) विभिन्न गैसों से।

16. भविष्य को सुन्दर बनाया जा सकता है
    (क) प्रदूषण रोककर। (ख) जनसंख्या बढ़ाकर।
    (ग) कारखाने लगाकर। (घ) कुछ भी न करके।

17. जनसंख्या वृद्धि का अर्थ है
    (क) निश्चित समय में बढ़ी लोगों की संख्या।
    (ख) संख्या का बढ़ना।
    (ग) जनसंख्या का लगातार बढ़ना।
    (घ) निश्चित वृद्धि।

18. जनसंख्या पर नियंत्रण आवश्यक है क्योंकि
    (क) उपलब्ध साधनों की संख्या सीमित है।
    (ख) उन्नति रुक जाएगी।
    (ग) निरक्षर अधिक होंगे।
    (घ) उपरोक्त सभी।

19. हमारे देश में जनगणना के लिए कौनसा ब्यौरा लिया जाता है
    (क) लोगों की संख्या का।
    (ख) परिवार के सदस्यों का, आयु का, शिक्षा का, व्यवसाय का, आय का।

(ग) केवल स्त्रियों का।
(घ) केवल पुरुषों का।

20. किसी देश की दो अलग-अलग वर्षों का तुलनात्मक रूप दर्शाने वाले ग्राफ को क्या कहेंगे?
(क) फोटोग्राफ (ख) ग्राफ
(ग) पिक्टोग्राफ (घ) तुलनात्मक ग्राफ

21. पिक्टोग्राफ लाभदायक है, क्योंकि
(क) आंकड़े बताता है।
(ख) अलग-अलग क्षेत्रों की स्थिति बताई जा सकती है।
(ग) आंकड़ों का तुलनात्मक अध्ययन हो सकता है।
(घ) उपरोक्त सभी।

22. जनसंख्या बढ़ने से पानी की कौन-सी समस्या उत्पन्न हो सकती है
(क) बीमार होने की सम्भावना का बढ़ जाना।
(ख) पीने के पानी की समस्याएं।
(ग) सिंचाई का प्रबन्ध न हो पाना।
(घ) उपरोक्त सभी।

23. मैदानी क्षेत्रों की जनसंख्या पहाड़ी क्षेत्रों की अपेक्षा
(क) अधिक होती है। (ख) कम होती है।
(ग) समान होती है। (घ) कोई अन्तर नहीं होता।

24. बंग्लादेश से निवासियों का उत्तर प्रदेश व बिहार में आना, जनसंख्या को
(क) प्रभावित नहीं कर रहा।
(ख) जनसंख्या कम हो रही है।
(ग) जनसंख्या बढ़ रही है।
(घ) लोगों में मेजलोज हो रहा है।

25. जनसंख्या बढ़ने से कौन-सा प्रदूषण होता है
(क) वायु प्रदूषण (ख) जल प्रदूषण
(ग) ध्वनि प्रदूषण (घ) उपरोक्त सभी

26. जनसंख्या बढ़ने से उद्योग-धन्धे भी
(क) कम हो रहे हैं। (ख) बढ़ रहे हैं
(ग) अधिक काम कर रहे हैं (घ) स्थान घेर रहे हैं।

27. क्या भारत में बाढ़ आने से जनसंख्या प्रभावित होती है?
    (क) हाँ (ख) नहीं
    (ग) कभी होती है, कभी नहीं होती (घ) उपरोक्त सभी।

28. नीचे दिए गए पिक्टोग्राफ (च), (छ) और (ज) में से कौन-सा पिक्टोग्राफ ज्यादा जनसंख्या दर्शा रहा है

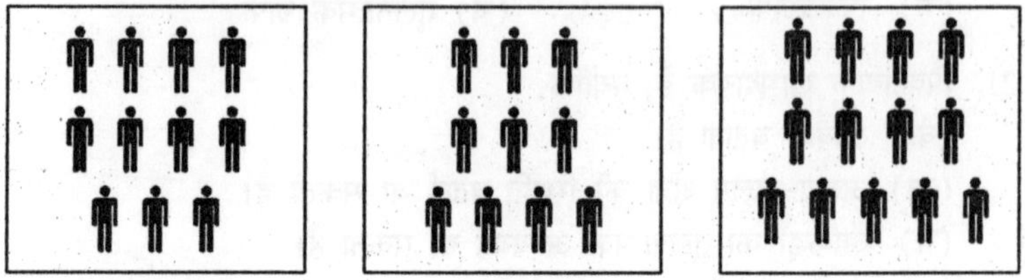

(च) (छ) (ज)

    (क) पिक्टोग्राफ (च)
    (ख) पिक्टोग्राफ (छ)
    (ग) पिक्टोग्राफ (ज)
    (घ) तीनों पिक्टोग्राफ बराबर की जनसंख्या दर्शा रहे हैं।

29. जनसंख्या वृद्धि समस्याएं उत्पन्न होती हैं
    (क) बेरोजगारी में वृद्धि।
    (ख) मूल्यों में निरन्तर वृद्धि।
    (ग) मकान व स्वास्थ्य संबंधी समस्या।
    (घ) उपरोक्त सभी।

30. मांसाहारी जन्तुओं के बढ़ने से
    (क) घास कम हो जाएगी।
    (ख) शाकाहारियों की संख्या कम हो जाएगी।
    (ग) शाकाहारी बढ़ जाएंगे।
    (घ) जन्तु बढ़ जाएंगे।

APPENDIX E

Group Test of General Mental Ability

मानसिक योग्यता की सामूहिक परीक्षा

इस प्रश्न-पुस्तिका के सभी उत्तरों को केवल उत्तर-पत्र पर ही लिखना होगा। इस परीक्षा पुस्तिका पर कुछ लिखना या चिन्ह न बनाना चाहिए।

प्रारम्भिक आदेश

हम आपकी सामान्य मानसिक योग्यता की परीक्षा करना चाहते हैं। केवल 20 मिनट का समय है। आप के सामने 100 प्रश्न आयेंगे।

इस परीक्षा के आरम्भ होने से पहले इसमें दिए गए सब प्रकार के प्रश्नों और उनके उत्तर लिखने की विधि को उदाहरण देकर समझाया जायेगा। हमें आशा है कि आपको उचित सफलता मिलेगी। सभी प्रश्न साधारण भाषा में हैं। प्रत्येक प्रश्न के दोनों ओर प्रश्न की क्रमिक संख्या छपी है। प्राय: सभी प्रश्नों के कुछ संभव वैकल्पिक उत्तर भी दिए गए है। हर एक वैकल्पिक उत्तर की संख्या भी उसके साथ छपी है। आपको हर प्रश्न को समझ कर केवल उसके सही उत्तर को चुनना है, तथा उस उत्तर की संख्या को तत्काल उत्तर पत्र के क्रम अनुसार उचित स्थान पर लिखना है। प्रत्येक प्रश्न का उत्तर संख्या में देना है। अर्थात् अक्षरों में कुछ नहीं लिखना है। ध्यान रखें प्रत्येक प्रश्न का एक ही ठीक उत्तर है। समय अधिक नहीं है। सब प्रश्नों का सही उत्तर बहुत कम लोग दे सकते हैं। अतएव आपको खूब शीघ्रता से काम करना चाहिए। और अधिक से अधिक सही प्रश्नों के उत्तर देने का प्रयास करना चाहिए। अगर कोई प्रश्न आपको अधिक कठिन लगता है, तब उस पर सोच विचार में अधिक समय नष्ट न करें। उसे छोड़ दें और उत्तर पत्र के निश्चित स्थान पर एक कोने में हल्का सा चिन्ह बना दें, और अगले प्रश्न का उत्तर सोच कर तुरन्त उसके उचित स्थान पर लिखें। यदि अन्त में समय हो, तो अपने उत्तरों को दोहरा लीजिए तथा छूटे हुए प्रश्नों का हल सोच कर लिखिए।

आरम्भ करने की आज्ञा सुनकर ही आप प्रश्नों को पढ़ने और उत्तर लिखने का कार्य आरम्भ करें, और जितनी शीघ्रता से ही सकें साफ उत्तर लिखिए।

एक बात और ध्यान रखिए इस प्रश्न पुस्तिका पर आपको कुछ नहीं लिखना है, और न इस पर किसी प्रकार का चिन्ह ही लगाना है।

केवल उत्तर-पत्र पर यथोचित स्थान में उत्तर की संख्या ही लिखनी है।

अभ्यास के लिए उदाहरण

इस परीक्षा में जिस प्रकार के प्रश्न पूछे गये हैं, उन के दो-दो उदाहरण नीचे दिए गये हैं। इनमें से पहले का उत्तर भी उत्तर-पत्र पर छपा है। किन्तु दूसरे का उचित उत्तर देने का अभ्यास आप सरलता से कर सकेंगे।

आइए अब हम इनको पढ़ें, और इनको हल करने की विधि समझें:

1. वृक्ष का अर्थ है,
   (1) पेड़ (2) जमीन
   (3) घास (4) फल

2. आज्ञा का अर्थ है,
   (1) कठोर (2) स्वामी
   (3) निर्देश (4) पालन

3. अच्छाई का उल्टा है
   (1) चालाकी (2) बुराई
   (3) लड़ाई (4) नम्रता

4. जीवन का उलटा है,
   (1) निराशा (2) आनन्द
   (3) मिट्टी (4) मृत्यु

5. नीचे दिए संख्या क्रम के अनुसार आगे की एक संख्या उत्तर-पत्र पर लिखें:
   1, 2, 3, 4, 5, 6, ...

6. नीचे दिए संख्या क्रम के अनुसार आगे की एक संख्या उत्तर-पत्र पर लिखें:
   15, 14, 13, 12, 11, 10

7. इन पांच शब्दों में से बेमेल शब्द की संख्या उत्तर-पत्र पर लिखें
   (1) घोड़ा, (2) मुर्गा, (3) हाथी, (4) मोर, (5) लड़का

8. इन पांच शब्दों में से बेमेल शब्द की संख्या उत्तर-पत्र पर लिखें
   (1) निबन्ध, (2) लेखक, (3) उपन्यास, (4) कविता, (5) स्तम्भ

9. छाता एक लाभदायक वस्तु है, इसलिए कि वह
   (1) कपड़े का बनता है।

(2) हमें धूप व वर्षा से बचाता है।
(3) वह सब देशों में मिलता है।

10. लोग बिल्लियां इसलिए पालते हैं, कि
(1) उनकी खाल कोमल होती हैं।
(2) वे कुत्तों से डरती हैं।
(3) वे चूहे पकड़ती हैं।

11. कलम : लिखना :: चाकू :
(1) आम, (2) काटना,
(3) लोहा, (4) खाना।

12. खीर : चावल :: हलवा :
(1) पूरी, (2) दही,
(3) दूध, (4) सूजी।

13. हरदेव से सुरजीत लम्बा है, किन्तु हरदेव से जगजीत नाटा है, तो सबसे लंबा कौन है?
(1) हरदेव, (2) सुरजीत, (3) जगजीत

14. राम के पीछे गोविन्द खड़ा है, गोबिन्द के पीछे चन्दन खड़ा है, और हरि के पीछे चन्दन खड़ा है, तो सबसे पीछे कौन खड़ा है?
(1) राम, (2) गोबिन्द,
(3) चन्दन, (4) हरि।

परीक्षा आरम्भ होने से पहले अपनी सभी शंकायें पूछ लीजिये।

मानसिक योग्यता की सामूहिक परीक्षा (72)

1. प्रकाश का उल्टा है,
   (1) काला, (2) लैम्प,
   (3) सूट, (4) अन्धकार।

2. कृपा का अर्थ है
   (1) धर्म, (2) कर्म,
   (3) दया, (4) दान।

3. पुलिस थाना चौबीस घण्टे खुला रहता है, क्योंकि
   (1) पुलिस अधिकारियों को 24 घण्टे का वेतन मिलता है।
   (2) लूटमार और दंगे की घटनाएं किसी भी समय हो सकती हैं।
   (3) पुलिस वालों को दिन-रात की वरदियां मिलती हैं।

4. मोटा का उल्टा है
   (1) छोटा, (2) पतला,
   (3) हल्का, (4) परिश्रमी।

5. घर का अर्थ है
   (1) बीवी, (2) परिवार,
   (3) मकान, (4) खर्च।

6. इन पांच शब्दों में से बे-मेल शब्द का अंक उत्तर पत्र पर लिखें:
   (1) प्लेट, (2) चम्मचा, (3) प्याला, (4) पतीला, (5) केला।

7. भीतर का उल्टा है
   (1) बाहर, (2) खुला,
   (3) मैदान, (4) तीव्र।

8. विद्या का अर्थ है,
   (1) पुस्तक, (2) ज्ञान,
   (3) रहस्य, (4) विज्ञान।

9. लक्ष्मण से आयु में सीता बड़ी है, परन्तु लक्ष्मण से भरत छोटा है। तब इनमें सबसे बड़ा कौन है?
   (1) लक्ष्मण (2) भरत (3) सीता

10. साधु का उल्टा है
    (1) झगड़ा, (2) दुष्ट,
    (3) भजन, (4) लड़का।

11. इन पांच शब्दों में से बेमेल शब्द का अंक उत्तर-पत्र पर लिखें:
    (1) मोटर, (2) साइकल, (3) तांगा, (4) तार, (5) रेलगाड़ी।

12. विष का उल्टा है
    (1) मीठा, (2) औषध,
    (3) अमृत, (4) शिव।

13. अहमद से अनवर नाटा है, किन्तु अनवर से हमीद नाटा है, तो सब से नाटा कौन है?
    (1) अहमद, (2) हमीद, (3) अनवर।

14. बलवान का अर्थ है
    (1) मोटा, (2) धनवान (3) प्रधान, (4) शक्तिमान।

15. इन पांच शब्दों में से बेमेल शब्द का अंक उत्तर-पत्र पर लिखें
    (1) चम्पा, (2) चमेली, (3) चाय, (4) गेंदा, (5) गुलाब।

16. अर्जुन से कमला अधिक दौड़ती है, किन्तु चपला से कमला पीछे रह जाती है, तो सबसे अधिक तेज़ कौन दौड़ता है
    (1) चपला, (2) कमला, (3) अर्जुन।

17. इन पांच शब्दों में से बेमेल शब्द का अंक उत्तर-पत्र पर लिखें।
    (1) मोटर, (2) रिक्शा, (3) तांगा, (4) पैदल, (5) साईकल।

18. "मुख में राम बगल में छुरी" का अभिप्राय: है
    (1) राम-राम कहने वाले सदा बगल में छुरी रखते हैं।
    (2) राम कहने से छुरी से रक्षा होती है।
    (3) अनेक दुष्ट लोग धर्म का पाखण्ड करते हैं।

19. नीचे दिए संख्या-क्रम के अनुसार आगे की एक संख्या उत्तर-पत्र पर लिखें
    8, 7, 6, 5, 4, 3, ...

20. इन पांच शब्दों में से बेमेल शब्द का अंक उत्तर-पत्र पर लिखें।
    (1) हृदय, (2) आँख, (3) कान, (4) नाक, (5) जीभ।

21. जूते-चमड़े के इसलिए बनते हैं कि
    (1) यह अधिक चलता है।

(2) वह मृत पशु की खाल से बनता है।
(3) यह सब देशों में पाया जाता है।

22. नीचे दिए संख्या क्रम के अनुसार आगे की एक संख्या उत्तर-पत्र पर लिखें
6, 11, 16, 21, 26, ...

23. सांच को आंच नहीं होती, इसलिए कहते हैं कि
(1) सच बोलने वाले को आग नहीं जलाती।
(2) सच की विजय होती है।
(3) सच्चे आदमी के घर में आंच नहीं मिलती।

24. नीचे दिए संख्या-क्रम के अनुसार आगे की एक संख्या उत्तर-पत्र पर लिखें
3, 6, 9, 12, 15, 18, ...

25. विदेश जाने के लिए लोग विमान यात्रा पसंद करते हैं इसलिए कि
(1) इसमें थोड़ा समय लगता है।
(2) यात्रा में खाने पीने का पूरा प्रबन्ध होता है।
(3) वह हवा में धूल से ऊपर उड़ते हैं।

26. इन पांच शब्दों में से बेमेल शब्द का अंक उत्तर-पत्र पर लिखें
(1) हाकी, (2) फुटबाल, (3) शतरंज, (4) क्रिकेट, (5) टेनिस।

27. पापी का मन सदा शंकित रहता है, इसलिए कि
(1) उसको नरक का कष्ट भोगना पड़ेगा।
(2) शंकित मन वाले पाप करते हैं।
(3) पापी को पोल खुलने का डर रहता है।

28. नीचे दिए संख्या क्रम के अनुसार आगे की एक संख्या उत्तर-पत्र पर लिखें
5, 11, 17, 23, 29, 35, ...

29. एक देश में रेल की बहुत सी लाइनें होनी चाहिए, इसलिए कि
(1) इनसे माल और मनुष्यों के आने जाने में सुविधा होती है।
(2) इनसे व्यापार को लाभ होता है।
(3) इनके द्वारा देश में खाद्य पदार्थों का मूल्य कम हो जाता है।

30. हीरा का अर्थ है
(1) मोती, (2) महंगा,
(3) पत्थर, (4) जवाहर।

31. नीचे दिए संख्या क्रम के अनुसार आगे की एक संख्या उत्तर-पत्र पर लिखें
3, 12, 21, 30, 39, 48, ...

32. इन पांच शब्दों में से बेमेल शब्द का अंक उत्तर-पत्र पर लिखें
    (1) कालीदास, (2) तुलसीदास, (3) जयशंकर प्रसाद, (4) बुद्ध, (5) टैगोर।

33. नीचे दिए संख्या क्रम के अनुसार की एक संख्या उत्तर-पत्र पर लिखें
    14, 17, 20, 23, 26, 29, ...

34. घोड़ा : टांग : गाड़ी :
    (1) बालक, (2) पहिया,
    (3) सड़क, (4) टट्टू।

35. इन पांच शब्दों में से बेमेल शब्द का अंक उत्तर-पत्र पर लिखें।
    (1) पास, (2) दूर, (3) परे, (4) यहां, (5) धीमा।

36. लिपिक : अध्यक्ष :: सैनिक :
    (1) मजदूर, (2) विक्रेता,
    (3) कप्तान, (4) चालक।

37. इन पांच शब्दों में से बेमेल शब्द का अंक उत्तर-पत्र पर लिखें।
    (1) खेलना, (2) सोना, (3) गाना, (4) दौड़ना, (5) नाचना।

38. तरल : ठोस :: पानी
    (1) बरफ, (2) मछली, (3) तैरना, (4) स्नान।

39. नीचे दिए संख्या क्रम के अनुसार की एक संख्या उत्तर-पत्र पर लिखें
    1, 2, 4, 8, 16, 32, ...

40. जनवरी : फरवरी :: जुलाई :
    (1) मार्च, (2) अगस्त,
    (3) रविवार (4) जून।

41. नीचे दिए संख्या क्रम के अनुसार आगे की एक संख्या उत्तर-पत्र पर लिखें
    21, 19, 17, 15, 13, 11, ...

42. बहन : भाई :: मासी :
    (1) चाचा, (2) बुआ,
    (3) दादा, (4) मामा।

43. गोवर्द्धन की मोटाई चन्द्रन से कम है, और चन्द्रन से अधिक मोटा गिरधारी है, तो सबसे दुबला कौन है?
    (1) चन्द्रन, (2) गिरधारी, (3) गोवर्द्धन।

44. नीचे दिए संख्या क्रम के अनुसार आगे की एक संख्या उत्तर-पत्र पर लिखें
18, 16, 14, 12, 10, 8, ...

45. हंसना : रोना :: बचपन :
(1) खेलकूद, (2) बुढ़ापा,
(3) मारपीट, (4) हार।

46. इन पांच शब्दों में से बेमेल शब्द का अंक उत्तर-पत्र पर लिखें।
(1) गाय, (2) भैंस, (3) घोड़ा, (4) भेड़, (5) बकरी।

47. क्रूर का उल्टा है
(1) सज्जन, (2) भला,
(3) दयालु, (4) कठोर।

48. इन पांच शब्दों में से बेमेल शब्द का अंक उत्तर-पत्र पर लिखें।
(1) कूदना, (2) फांदना, (3) भागना, (4) खड़े रहना, (5) चलना।

50. इन पांच शब्दों में से बेमेल शब्द का अंक उत्तर-पत्र पर लिखें।
(1) मिट्टी, (2) लकड़ी, (3) शिला, (4) कंकर, (5) पत्थर।

51. उद्यम का उल्टा है
(1) वियोग, (2) डरपोक,
(3) विश्राम, (4) आलस्य।

52. नीचे दिए संख्या क्रम के अनुसार आगे की एक संख्या उत्तर-पत्र पर लिखें
78, 67, 56, 45, 34, 23, ...

53. फल : सेब :: पुष्प :
(1) अनार, (2) बादाम,
(3) गुलाब, (4) जामुन।

54. मोहन से नाटा राम है। और किशन से नाटा राम है। तब सबसे कम लम्बा कौन है?
(1) मोहन, (2) किशन, (3) राम।

55. नीचे दिए संख्या क्रम के अनुसार आगे की एक संख्या उत्तर-पत्र पर लिखें
5, 6, 8, 11, 15, 20, ...

56. "झूठ के पांव नहीं होते" यह इस कारण कहा जाता है कि
(1) लंगड़े मनुष्य बहुत झूठ बोलते हैं।

(2) झूठे मनुष्य की पोल शीघ्र खुल जाया करती है।
(3) झूठ बोलने वाले बहुत बार चलते समय ठोकर खाते हैं।

57. नाव : माझी :: मोटर :
(1) स्वामी, (2) यात्री,
(3) नगर, (4) चालक।

58. इन पांच शब्दों में से बेमेल शब्द का अंक उत्तर-पत्र पर लिखें।
(1) खाट, (2) कुर्सी, (3) प्लेट, (4) सोफा, (5) पीढ़ा।

59. मकान : ईंट :: सेना :
(1) सिपाही, (2) पत्थर,
(3) हथियार, (4) युद्ध।

60. नीचे दिए संख्या क्रम के अनुसार आगे की एक संख्या उत्तर-पत्र पर लिखें
5, 6, 9, 10, 13, 14, ...

61. संपादक : पत्रिका :: व्यापारी :
(1) बाजार, (2) विज्ञापन,
(3) दुकान, (4) समाचार।

62. आश्चर्य का अर्थ है,
(1) निराला, (2) विस्मय,
(3) घबराहट, (4) अनुभव।

63. चन्द्रमा : पृथ्वी :: पृथ्वी :
(1) सागर, (2) मंगलतारा,
(3) सूर्य, (4) मछलियां।

64. इन पांच शब्दों में से बेमेल शब्द का संख्या उत्तर-पत्र पर लिखें।
(1) गोभी, (2) गाजर, (3) कक्कड़ी, (4) मूली, (5) धनिया।

65. सोना का अर्थ है
(1) कनक, (2) खाद, (3) घन, (4) माला।

66. नीचे दिए संख्या क्रम के अनुसार आगे की एक संख्या उत्तर-पत्र पर लिखें
9, 12, 14, 17, 19, 22, ...

67. इन पांच शब्दों में से बेमेल शब्द का अंक उत्तर-पत्र पर लिखें।
(1) भूखा, (2) मरा, (3) प्यासा, (4) थका, (5) हारा।

68. सदाशिव से मुरारी लम्बा है। किन्तु मुरारी से वीरेन्द्र नाटा है। और त्रिलोकी से मुरारी नाटा है, तो सबसे लम्बा कौन है?
(1) सदाशिव, (2) मुरारी,
(3) वीरेन्द्र, (4) त्रिलोकी।

69. वृक्ष : लता :: फल :
(1) फूल, (2) चम्पा,
(3) मोतिया, (4) मौलसिरी।

70. नीचे दिए संख्या क्रम के अनुसार आगे की एक संख्या उत्तर-पत्र पर लिखें
8, 9, 12, 13, 16, 17, ...

71. नेता : जनता :: अधिकारी :
(1) चुनाव, (2) भाषण,
(3) कर्मचारी, (4) निर्णय।

72. आरेखन-कला में राम से गार्गी चतुर है। किन्तु उसकी अपेक्षा सीता चतुर है। अतः सबसे चतुर कौन है?
(1) गार्गी, (2) सीता, (3) राम।

73. इन पांच शब्दों में से बेमेल शब्द का अंक उत्तर-पत्र पर लिखें।
(1) घोड़ा, (2) ऊंट, (3) कंगारू, (4) गधा, (5) भैंसा।

74. चित्र : खड़ा :: सिनेमा :
(1) खाता, (2) चलता,
(3) हंसता, (4) रोता।

75. नीचे दिए संख्या क्रम के अनुसार आगे की एक संख्या उत्तर-पत्र पर लिखें
29, 28, 26, 23, 19, 14, ...

76. मेरे विचार में यदु से सीता चतुर है। किन्तु कला से रमा निःसन्देह चतुर है। और सीता से रमा मन्द है। तो सबसे चतुर कौन है।
(1) यदु, (2) कमला, (3) रमा, (4) सीता।

77. नीचे दिए संख्या क्रम के अनुसार आगे की एक संख्या उत्तर-पत्र पर लिखें
7, 8, 10, 13, 17, 22, ...

78. सभ्यता का अर्थ है,
(1) वस्त्र, (2) कला,
(3) विज्ञान, (4) संस्कृति।

79. "जिसकी लाठी उसकी भैंस" कहने का अभिप्राय है कि
    (1) भैंस वाले के पास लाठी आवश्यक होती है।
    (2) अधिक बलवान की बात सबको माननी पड़ती है।
    (3) लाठी देखकर भैंस अधिक दूध देती है।

80. जूलियट : रोमियो :: संयोगिता :
    (1) स्वयंवर, (2) जयचन्द,
    (3) पृथ्वीराज, (4) अकबर।

81. अनेक वर्षों तक हवाई जहाज सफल न हुए, क्योंकि
    (1) वे बहुत भारी बनाए जाते थे।
    (2) उनके कल पुर्जे बहुत जटिल होते थे।
    (3) एक उत्तम इंजन नहीं बन पाया था।

82. नीचे दिए संख्या क्रम के अनुसार आगे की एक संख्या उत्तर-पत्र पर लिखें
    4, 6, 9, 11, 14, 16, ...

83. इन पांच शब्दों में से बेमेल शब्द का अंक उत्तर-पत्र पर लिखें
    (1) चिड़िया, (2) तोता, (3) बुलबुल, (4) कबूतर, (5) उल्लू।

84. नीचे दिए संख्या क्रम के अनुसार आगे की एक संख्या उत्तर-पत्र पर लिखें
    8, 9, 11, 12, 14, 15, ...

85. इन पांच शब्दों में से बेमेल शब्द का अंक उत्तर-पत्र पर लिखें।
    (1) गया, (2) पुरी, (3) प्रयाग, (4) द्वारिका, (5) दिल्ली।

86. दैनिक : मासिक :: पत्र :
    (1) कहानियाँ, (2) समाचार,
    (3) पत्रिका, (4) संवाद।

87. ऋण का उल्टा है
    (1) धन, (2) बचत, (3) बनिया, (4) व्यापार।

88. कोट : पैन्ट :: कुरता :
    (1) समाचार, (2) टोपी,
    (3) पाजामा, (4) पगड़ी।

89. विस्तृत का उल्टा है
    (1) विशाल, (2) कमरा,
    (3) पतला, (4) संकुचित।

90. पेनसिल : चाक :: कापी :

(1) पुस्तक, (2) बोर्ड,

(3) ताक, (4) लेख।

91. इन पांच शब्दों में से बेमेल शब्द का अंक उत्तर-पत्र पर लिखें।

(1) पहर, (2) प्रभाव, (3) घंटा, (4) मिनट, (5) क्षण।

92. नीचे दिए संख्या क्रम के अनुसार आगे की एक संख्या उत्तर-पत्र पर लिखें

2, 3, 5, 6, 8, 9, ...

93. श्वेत : हिम :: श्याम :

(1) दिन, (2) चिड़िया,

(3) कौआ, (4) रात।

94. नीचे दिए संख्या क्रम के अनुसार आगे की एक संख्या उत्तर-पत्र पर लिखें

27, 26, 24, 21, 17, 12, ...

95. विहंग का अर्थ है,

(1) मोर, (2) पक्षी,

(3) भद्दा, (4) निषंग।

96. राम की बुद्धि देवकी से प्रखर है, पर सीता की बुद्धि सावित्री से हीन है, किन्तु देवकी की बुद्धि सावित्री से उत्तम है, तो सबसे बुद्धिमान कौन है?

(1) देवकी, (2) सीता,

(3) रमा, (4) सावित्री।

97. इन पांच शब्दों में से बेमेल शब्द का अंक उत्तर-पत्र पर लिखें

(1) हिमाचल, (2) केरल, (3) मेघालय, (4) भोपाल, (5) हरियाणा।

98. नाटा का उल्टा है

(1) भारी, (2) लम्बा, (3) तगड़ा, (4) कठोर।

99. गरल का अर्थ है

(1) बुरा, (2) विष,

(3) कड़वा, (4) सरल।

100. मकड़ी : मक्खी :: बिल्ली :

(1) कुत्ता, (2) पिल्ला,

(3) दूध, (4) चूहा।

(यदि समय बाकी है तो अपने उत्तरों को दोहराइये)

APPENDIX F

पाठ योजना विज्ञान शिक्षण

Lesson plan for the experimental group

पाठ योजना

विषय-विज्ञान
उपविषय-जनसंख्या
(परिचय व जनसंख्या की गणना)

तिथि -7/1/98
कक्षा -VII अ
कालांश-पहला
अवधि-40 मिनट

व्यवहारगत उद्देश्य

- अपने परिवेश के संदर्भ में विद्यार्थी जनसंख्या की परिभाषा से परिचित होंगे।
- विद्यार्थी छोटे परिवार व बड़े परितार के अंतर को समझेंगे। वे छोटे परिवार के प्रति एक सकारात्मक दृष्टिकोण अपनाएँगे।
- विद्यार्थी अपने देश में जनसंख्या वृद्धि के दूरदर्शन, रेडियो व अखबार में आने वाले संदेशों के महत्व को समझेंगे।
- विद्यार्थी खेल द्वारा समझ राबेंगे की समय के साथ जनसंख्या में किस प्रकार बदलाव आता है।
- विद्यार्थी अपने मोहल्ले में अध्यापिका द्वारा करवाई गई जनसंख्या की गणना नामक क्रिया द्वारा समझ सकेंगे कि जनसंख्या की गणना के लिए किन-किन बातों का ब्यौरा तैयार करते हैं।

पूर्वज्ञान

बालक को जनसंख्या के बारे में सामान्य जानकारी है।

प्रस्तावना

- सोनू तुम्हारे परिवार में कितने सदस्य हैं?
- मनीष तुम्हारे परिवार की सदस्य संख्या क्या है?

– बच्चों, क्या हम एक परिवार के सदस्यों की संख्या को जनसंख्या कह सकते हैं?
– तो जनसंख्या क्या होती है? (समस्यात्मक प्रश्न)

छोटा परिवार

बड़ा परिवार

<table>
<tr><td>

श्यामपट्ट कार्य

विषय: विज्ञान — कक्षा-VII अ

उपविषय: जनसंख्या परिचय व गणना — कालाशं-पहला

अवधि-40 मिनट

जनसंख्या: किसी स्थान विशेष में रहने वाले लोगों (पुरुष, स्त्री व बच्चों) की संख्या जनसंख्या होती है।

जनगणना का ब्यौरा:

- परिवार के सदस्यों की संख्या
- उनकी आयु
- उनकी शिक्षा
- उनकी भाषा
- उनका धर्म
- उनका व्यवसाय
- उनकी आय

</td></tr>
</table>

उद्देश्य कथन

आज हम जनसंख्या के बारे में जानकारी प्राप्त करेंगे।

प्रस्तुतीकरण

शिक्षण बिन्दु	*अध्यापिका क्रियाएँ*	*विद्यार्थी क्रियाएँ*
जनसंख्या से अभिप्राय	बच्चों, जब हम कहते हैं कि कुरूक्षेत्र की अपेक्षा अम्बाला की जनसंख्या अधिक है1 इससे हमारा अभिप्राय क्या है? इससे हमारा अभिप्राय वहाँ रहने वाले लोगों-पुरुष, स्त्रियों व बच्चों की संख्या से होता है। अर्थात् किसी स्थान की जनसंख्या का अर्थ एक ही जाति (स्पेसीज) के उन जीवों से होता हैं जो एक निश्चित समय तक एक साथ रहते हैं।	विद्यार्थी ध्यानपूर्वक अध्यापिका की बातें सुनेंगे तथा पूछे गए प्रश्नों का उत्तर देने की कोशिश करेंगे।
छोटा परिवार एवं बड़ा परिवार	कक्षा को चाक से चार बराबर भागों में विभाजित किया जाएगा। अब छात्रों को बड़े एवं छोटे परिवार का अभिनय करने के लिए चार परिवारों में बाटाँ जाएगा। दो बड़े	विद्यार्थी छोटे परिवार व बड़े परिवार में रहने

(क्रमशः)

(क्रमशः)

शिक्षण बिन्दु	अध्यापिका क्रियाएँ	विद्यार्थी क्रियाएँ
	-परिवार बनाए जाएँगे जिनमें माता-पिता व छः-सात बच्चे होंगे और दो छोटे परिवार बनाए जाएँगे जिनमें माता-पिता और 2 बच्चे थे। चारों परिवारों को कक्षा का एक-2 भाग दे दिया गया और अपनी जीविका चलाने के लिए कहा जाएगा। सभी परिवार को दो-दो सेब राशन के प्रतीक रूप में दिये जाएँगें और उन्हें आपस में बाँटने को कहा जाएगा। बड़े परिवार के बच्चों को खाने के लिए बहुत कम मात्रा में मिलेगा जबकि छोटे परिवार के लिए पर्याप्त मात्रा में भोजन मिलेगा। बड़े परिवार के बच्चों में खेती का बटवारा होने पर बहुत कम जमीन बटेगी। इस प्रकार बड़े परिवार को रहने के लिए भी जगह कम पड़ेगी। जबकि छोटे परिवार के लिए उतनी ही जगह पूरी पड़ेगी, इस प्रकार बच्चों को छोटा परिवार व बड़े परिवार का अन्तर समझाया जाएगा।	वाले सदस्यों का अभिनय करेंगे।
संचार माध्यमों के संदेशों को समझना।	अध्यापिका सरकार द्वारा विभिन्न संचार माध्यमों से प्रशारित संदेशों की ओर छात्रों का धयान आकर्षित करेगी।	विद्यार्थी विभिन्न समाचार पत्रों व पत्रिकाओं से जनसंख्या संबंधी सूचनांए एकत्रित करेंगे
जनसंख्या की वृद्धि	इस खेल द्वारा विद्यार्थी समझ सकेंगे कि समय के साथ जनसंख्या में किस प्रकार बदलाव आता है1 विद्यार्थियों को दो समूहों में विभाजित किया जाएगा। एक समूह का नाम जनक व दूसरे समूह का नाम मृत्युदाता रखा जाएगा। अब कक्षा में 10-10 पत्थरों के पाँच समूह बनाए। जनक समूह के बच्चों को हर एक मिनट पश्चात् प्रत्येक पत्थर समूह में पाँच पत्थर और डालने के लिए कहा जाएगा। मृत्युदाता समूह के बच्चों को इसके पश्चार्त् प्रत्येक समूह में से दो पत्थर निकाल लेने होंगे। इस प्रकार यह क्रिया दस बार दोहराई जाएगी। बच्चों को खेल के अंत में प्रत्येक पत्थर समूह के अतिरिक्त पत्थरों व निकाले गए पत्थरों को गिनने के लिए कहा जाएगा।	विद्यार्थी अध्यापिका द्वारा करवाई गई क्रिया में भाग लेंगे1

(क्रमशः)

(क्रमशः)

शिक्षण बिन्दु	*अध्यापिका क्रियाएँ*	*विद्यार्थी क्रियाएँ*
	हर मिनट पश्चात् प्रत्येक समूह के पत्थरों की संख्या कितनी बढ़ी व खेल के पश्चात् पत्थरों की कुल संख्या कितनी बढ़ी यह गणना की जाएगी। कुछ समय उपरांत जनसंख्या वृद्धि किस प्रकार होती है इन तथ्यों से संबंधित जानकारी दी जाएगी।	
जनसंख्या की गणना	हमें कैसे पता चलता है कि जनसंख्या बढ़ रही है या घट रही है? जनसंख्या की गणना द्वारा हमें जनसंख्या के घटने बढ़ने का पता चलता है। जनगणना कैसे होती है? यह जानने के लिए हम खेल खेलेंगे। अध्यापिका पाँच डिब्बे लेगी जिनमें क्रमशः 9, 6, 12, 10, 14 अलग-अलग रंग व आकार के कंचे भरे हुए हैं।	सभी विद्यार्थी ध्यानपूर्वक अध्यापिका की बातें सुने व पूछे गए प्रश्न का उत्तर देने की कोशिश करेंगे।
खेल जनगणना कैसे होती है?	अब पाँच बच्चों को बुला कर सबको एक-एक डिब्बे के कंचे गिनने को कहेगी तथा एक कागज पर निम्न ब्यौरा लिखने को कहेगी: –कुल कंचों की संख्या –कुल बड़े कंचों की संख्या –कुल छोटे कंचों की संख्या –कुछ मोटे कंचों की संख्या –लाल रंग के कंचों की संख्या –नीले रंग के कंचों की संख्या ऐसे ही पांचों छात्रों को यह ब्यौरा लिखने को देगी। इसके बाद सभी कंचों को एक डिब्बे में डालकर एक अन्य विद्यार्थी को बुलाकर उससे कहेगी कि बताओ कि किस डिब्बे में किस रंग के व किस आकार के कितने कंचे थे? फिर सभी छात्रों से यही सवाल पूछेगी। बच्चे नहीं बता पाएंगे। अध्यापिका उनको बताएगी कि इस तरह इक्ट्ठे करके यह जानकारी प्राप्त करना कठिन है। इसी प्रकार जनगणना करते समय हमें जनसंख्या को विभिन्न भागो जैसे किसी भी राज्य की जनसंख्या की गणना के लिए उस राज्य के प्रत्येक गांव, कस्बे शहर की जनसंख्या की गणना अलग-अलग की जाती है। यदि पूरे देश	विद्यार्थी खेल में भाग लेंगे तथा अन्य सभी विद्यार्थी ध्यानपूर्वक देखेंगे।

(क्रमशः)

(क्रमशः)

शिक्षण बिन्दु	*अध्यापिका क्रियाएँ*	*विद्यार्थी क्रियाएँ*
	की जनसंख्या की गणना करनी हो तो इसी प्रकार अलग-अलग राज्यों की जनगणना करके उसका योग कर दिया जाता है। इसके लिए योजना कार्यालय के जनसंख्या विभाग के कर्मचारी घर-घर जाकर उसमें रहने वाले स्त्री, पुरुष व बच्चों के संबंध में अनेक प्रकार की जानकारी प्राप्त करते हैं।	
जनसंख्या से संबंधित ब्यौरा	अध्यापिका पूछेगी कि क्या आपके घर में कभी कोई जनसंख्या से संबंधित जानकारी या पूछताछ करने आया है? किसी भी देश या राज्य की जनगणना के लिए निम्न बातों का ब्यौरा तैयार किया जाता है: परिवार के सदस्यों (स्त्री, पुरुष और बच्चों) की संख्या-उनकी आयु-उनकी शिक्षा-उनकी भाषा-उनका धार्म-उनका व्यवसाय-उनकी आय आदि। विभिन्न प्रान्तों की जनसंख्या की गणना का कुल योग ही किसी देश की जनसंख्या होती हैं।	सभी विद्यार्थी ध्यानपूर्वक अध्यापिका द्वारा पूछे गये प्रश्नों को सुनकर उत्तर देने की कोशिश करेंगे।
स्वम् ब्यौरा तैयार करना	क्रिया द्वारा विद्यार्थी जनसंख्या की गणना के लिए वाँछित आँकड़ों व ब्यौरा तैयार करने के विषय में जानकारी प्राप्त करेंगे। अध्यापिका विद्यार्थियों के दो समूह बनाकर उन्हें अपने मुहल्ले या अपने पड़ोस के मुहल्ले के किसी परिवार के विषय में निम्नलिखित ब्यौरा तैयार करने के लिए कहेंगी: 1. परिवार के कुल सदस्य स्त्रियाँ, पुरुष, बच्चे 2. उनकी आयु 3. उनकी शिक्षा 4. उनकी भाषा 5. उनका धर्म 6. उनका व्यवसाय 7. उनकी आय इस प्रकार विद्यार्थी कई परिवारों की जनसंख्या का ब्यौरा तैयार करेंगे।	विद्यार्थी ब्यौरा तैयार करेंगे।

पुनरावृत्ति–पुनरावृत्ति के लिए अध्यापिकाएँ प्रश्न पूछेंगी:
- जनसंख्या की गणना कौन करता है?
- जनसंख्या की गणना कैसे होती है?
- जनसंख्या की गणना में किन बातों का ब्यौरा तैयार किया जाता है?

गृहकार्य: सभी विद्यार्थी अपने मुहल्ले की जनसंख्या की गणना करेंगें।

Lesson plan for the control group

पाठ योजना

विषय-विज्ञान — तिथि -7/1/98
उपविषय-जनसंख्या — कक्षा -VII ब
(परिचय व जनसंख्या की गणना) — कालांश-चौथा
अवधि-40 मिनट

व्यवहारगत उद्देश्य

- विद्यार्थी जनसंख्या के अभिप्राय को समझ सकेंगे
- विद्यार्थी जनसंख्या की गणना कैसे करते हैं बता सकेंगे-विद्यार्थी जनसंख्या का ब्यौरा दे सकेंगे।

पूर्वज्ञान

बालक को जनसंख्या के बारे में सामान्य जानकारी हैं 1

प्रस्तावना

- सोनू तुम्हारे परिवार में कितने सदस्य हैं?
- मनीष तुम्हारे परिवार की सदस्य संख्या क्या है?
- बच्चों, क्या हम एक परिवार के सदस्यों की संख्या को जनसंख्या कह सकते हैं?
- तो जनसंख्या क्या होती है? (समस्यात्मक प्रश्न)

श्यामपट्ट कार्य

विषय: विज्ञान कक्षा: VII अ

उपविषय: जनसंख्या परिचय व गणना कालांश: पहला

अवधि: 40 मिनट

जनसंख्या: किसी स्थान विशेष में रहने वाले लोगों (पुरुष, स्त्री व बच्चें) की संख्या जनसंख्या होती है।

जनगणना का ब्यौरा

–परिवार के सदस्यों की संख्या
–उनकी आयु
–उनकी शिक्षा
–उनकी भाषा
–उनका धर्म
–उनका व्यवसाय
–उनकी आय

उद्देश्य कथन

आज हम जनसंख्या के बारे में जानकारी प्राप्त करेंगे।

प्रस्तुतीकरण

शिक्षण बिन्दु	*अध्यापिका क्रियाएँ*	*विद्यार्थी क्रियाएँ*
जनसंख्या से अभिप्राय	सर्वप्रथम अध्यापिका बच्चों से पाठ को बारी-बारी से पढ़ने के लिए कहेगी। पाठ पढने के बाद अध्यापिका छात्रों को जनसंख्या के अभिप्राय से अवगत करायेगी।	विद्यार्थी बारी-बारी से पाठ वाचन करेंगे
जनसंख्या की गणना	बच्चों, जब हम कहते हैं कि कुरूक्षेत्र की अपेक्षा अम्बाला की जनसंख्या अधिक है। इससे हमारा अभिप्राय क्या है? इससे हमारा अभिप्राय वहाँ रहने वाले लोगों— पुरुष, स्त्रियाँ व बच्चों की संख्या से होता है। अर्थात् किसी स्थान की जनसंख्या का अर्थ एक ही जाति (स्पेसीज) के उन जीवों से होता है जो एक निश्चित समय तक एक साथ रहते है। हमें कैसे पता चलता है कि जनसंख्या बढ़ रही है या घट रही है? जनसंख्या की गणना के द्वारा हमें जनसंख्या के घटने बढ़ने का पता चलता है। जनसंख्या की गणना कैसे होती है।	विद्यार्थी ध्यानपूर्वक अध्यापिका की बातें सुनेंगे तथा पूछे गए प्रश्नों का उत्तर देने की कोशिश करेंगे।

(*क्रमशः*)

(क्रमशः)

शिक्षण बिन्दु	*अध्यापिका क्रियाएँ*	*विद्यार्थी क्रियाएँ*
	किसी प्रान्त की जनसंख्या की गणना के लिए उसके प्रत्येक गांव, कस्बों तथा शहरों की जनसंख्या की गणना की जाती है। गणना कौन करता है ? इसके लिए योजना कार्यालय के जनसंख्या विभाग के कर्मचारी घर-घर जाकर उसमें रहने वाले स्त्री, पुरुष व बच्चों के संबंध में अनेक प्रकार की जानकारी प्राप्त करते हैं। अध्यापिका पूछेगी कि क्या आपके घर में कभी कोई जनसंख्या से संबंधित जानकारी या पूछताछ करने आया है?	
जनसंख्या का ब्यौरा	किसी भी देश या राज्य की जनगणना के लिए निम्न बातों का ब्यौरा तैयार किया जाता है: परिवार के सदस्यों (स्त्री, पुरुष, बच्चों) की संख्या –उनकी आयु –उनकी शिक्षा –उनकी भाषा –उनका धर्म –उनका व्यवसाय –उनकी आय आदि। विभिन्न प्रान्तों की जनसंख्या की गणना का कुल योग ही किसी देश की जनसंख्या होती है। यदि पूरे देश की गणना करनी हो तो इसी प्रकार अलग-अलग राज्यों की जनगणना करके उसका योग कर दिया जाता है।	सभी विद्यार्थी ध्यानपूर्वक अध्यापिका की बातें सुने व पूछे गए प्रश्न का उत्तर देने की कोशिश करेंगे।

पुनरावृत्ति–पुनरावृत्ति के लिए अध्यापिका प्रश्न पूछेंगी:
- जनसंख्या की गणना कौन करता है?
- जनसंख्या की गणना कैसे होती है?
- जनसंख्या की गणना में किन बातों का ब्यौरा तैयार किया जाता है?

गृहकार्य: सभी विद्यार्थी अपने मुहल्ले की जनसंख्या की गणना कीजिए।

APPENDIX G

Incorporations in General Presentation of the Textbooks

Source.

1. Natrajan, C. (1997). Activity Based Foundation Course on Science, Technology and Society, Curriculum Book-I, Homi Bhabha Centre for Science Education, Tata Institute of Fundamental Research.
2. Stannard, P. and Williamson, K. (1987). Exploring Science, Ecos Educational Publication, Australia.
3. Probing the Natural World (1980), Silver Burdett Company, New Jersey.

Mastary Test

1. What size of graduated cylinder would you use to measure out—
   5 mL?
   60 mL?
2. What are the readings on the graduates shown ?
3. When measuring volume, why do you hold the graduate so that meniscus is level with your eye?

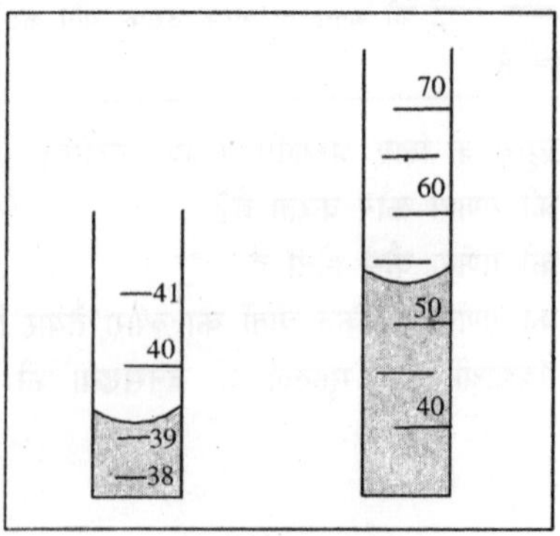

Accuracy of Graduate

4. Obtain a container (a cup, bottle or pill-bottle) from your teacher. Select a graduate which will measure most accurately the volume of the container.
   Fill in the container completely with water and then measure the volumes.
   Partner Check: Show the results to your partner, including the size of the graduate which you selected.

Your partner will repeat the measurement to check your result.

If there is a difference between your partner's reading and your reading, repeat the measurements.

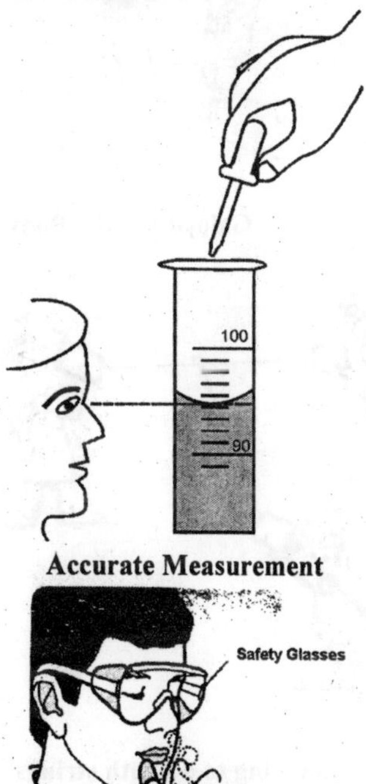

Accurate Measurement

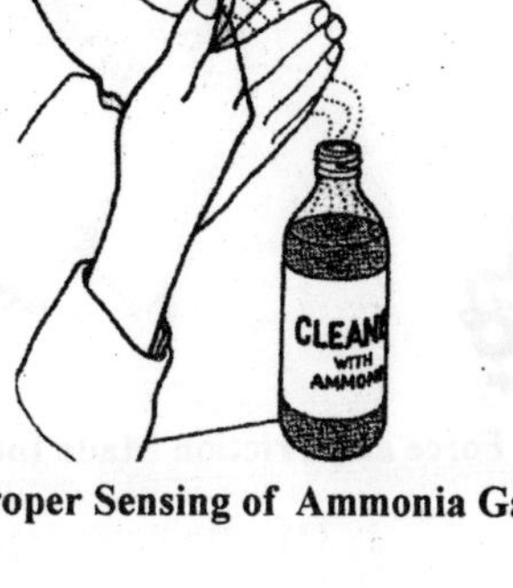

Proper Sensing of Ammonia Gas

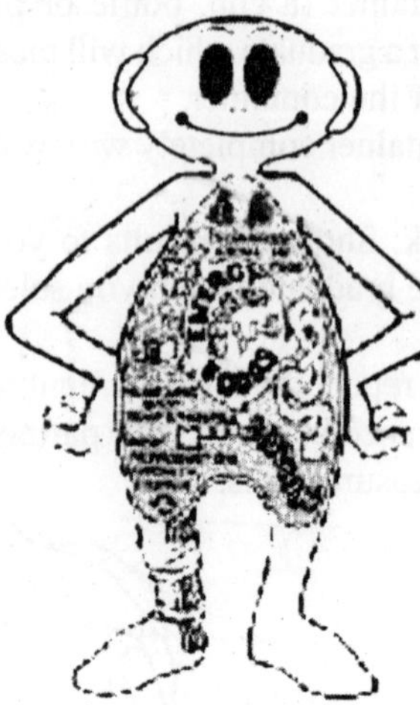

Calories in the Body

Something to do with strings

Force and Friction Made Interesting

POPULATION IN THE NEXT MILLENNIUM

With the Indian population having crossed the dreaded one billion mark in India's fifty third year of independence; it is imperative that we take a long hard serious look at our ever burgeoning population.

Population is one resource that never depletes and is a living development parameter. But is at times interpreted as a hindering factor for development. This happens because population is both a consumer and producer. There are two schools of thought. One which treats population as a resource and the other as a burden to the society. The truth, in fact, has somewhere in between. The interplay of factors responsible for population growth and those for development decide the resourcefulness of population. Let us see how the Indian population has grown over the last century.

Government was enable to control after independence epidemic diseases like small-pox, plague, malaria and so on

These developments were not accompanied by significant reduction in fertility and hence the 'population explosion' during the last 40 years. A slight reduction in the pace of growth during the last decade (1981-91) has created hope for lesser intensity of population explosion in future.

REGULATION OF REPRODUCTION

Although there are many individuals and organisation, which tried to propagate the idea of family planning, it was in 1951, that the Government of India for the first time adopted the family planning programme. The sub-committee on family planning strongly recommended. 15-20 years. The idea of stopping childbearing artificially was an idea hard to accept. The programme was hardly voluntary and stressed the need to provide family-planning solution at people's doorstep. There was a need to integrate them with medical and public health services, especially with maternal and child-health programmes.

During the mid-60s, efforts were made to push the programme, forcefully, through mass vasectomy camps. Although the numbers were indicative of the success, they brought with them the adverse consequences of neglect and carelessness in the mass camps and a bad name to vasectomy.

The government could never popularise the programme after that. In the third five-year plan, a goal of birth rate of 25 per 1000 was introduced in government centres/ hospitals in urban areas. The World Population Conference in Bucharest (1974) added a new dimension to the programme. 'Development is the best Contraceptive was massage of the day.'

The year 1976, the worst period, as far as governance of the nation and population policies were concerned came up with a new approach. Along with the declaration of 'Emergency', coercion and mild pressure were used in implementing the family planning programme. The central assistance of 8 per cent was linked with the family planning performance. For the first time the Union Government allowed some states to bring legislation for compulsory sterilization.

Newspaper Clipping

Comic Figures on Observation and Understanding of Communication

Concept of Detergent in Water Made Interesting

HOW TO BUILD A PINHOLE CAMERA

1. Cut the end out of a tin can punch a hole with a nail in the other. Painting the can black inside helps to form a better image.

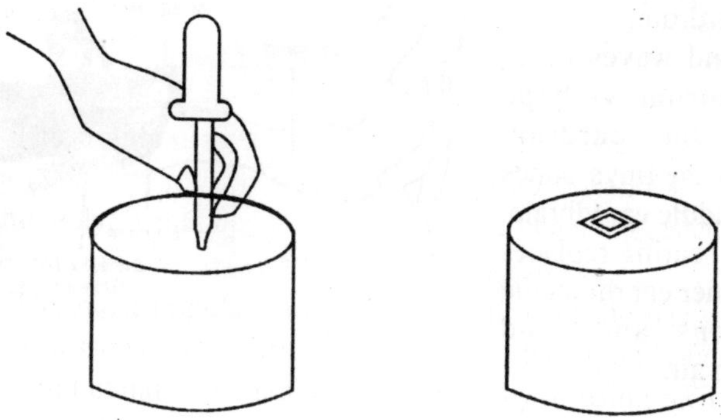

2. Cover the hole with a piece of aluminium foll, and punch a very tiny hole in the left.

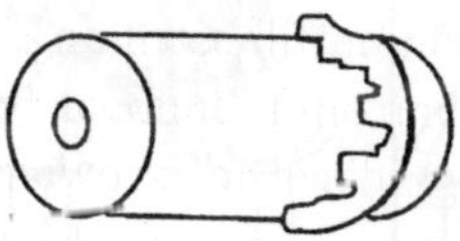

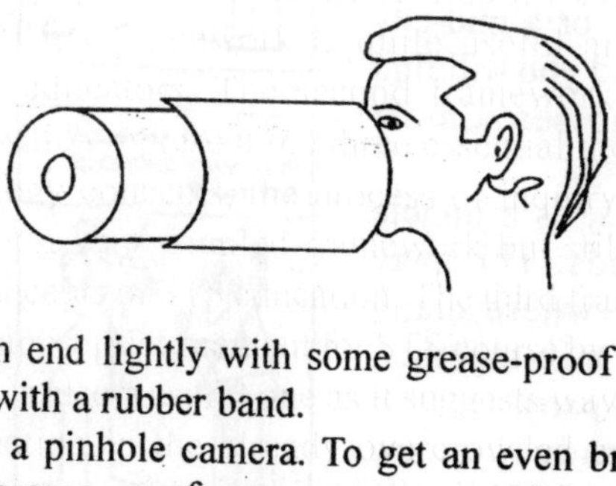

3. Cover the open end lightly with some grease-proof paper and hold it on with a rubber band.
4. You now have a pinhole camera. To get an even brighter image on the grease-proof paper, warp a cardboard tube around the tin.

Activity with Diagrammatic Representation and Its Conduction

HEARING

Sounds are detected by your ear. When you hear a sound, you turn your head so that your ear faces the sound. The outer line of your ear channels the sound into your ear. The sound then goes along the ear canal to the eardrum.

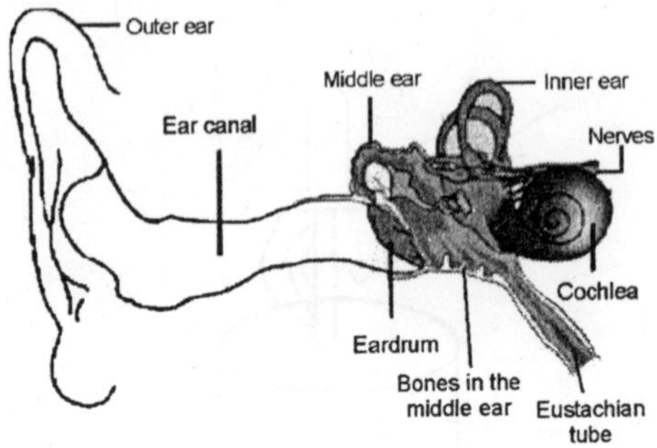

Sound waves make the eardrum v-shape finally the eardrum makes these tinys bones in the middle ear vibrate. Finally, hearing replaces in the inner ear direct the vibrations and send sound in air.

Noise Measurement

Some noises are pleasant. Others are very unpleasant. Some noises are both loud and unpleasent. Stand on the side of a busy highway or a mid-city street and you will find some noises almost deafening.

Noise is a jumble of sounds. This may happen where many sounds mix together, like at a party.

Sounds are measured in decibels.

Cartoon Presentation in this Chapter of Sound

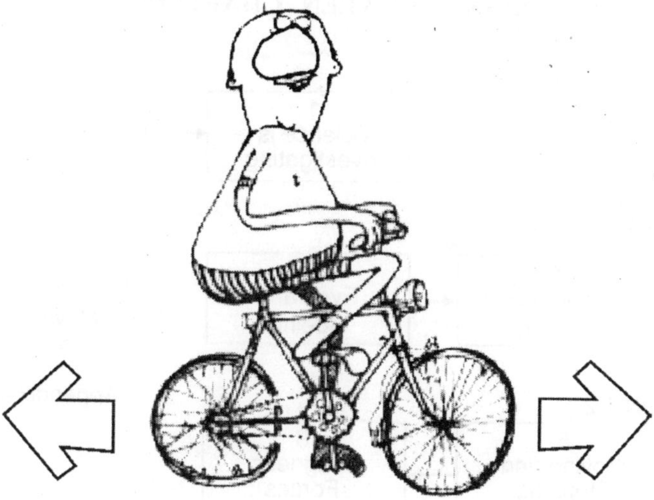

When you push harder on the pedals, the bicycle moves faster. This time, the forces are unbalanced.

Moving and Stopping
Balancing Forces

LINKS BETWEEN CHAPTERS

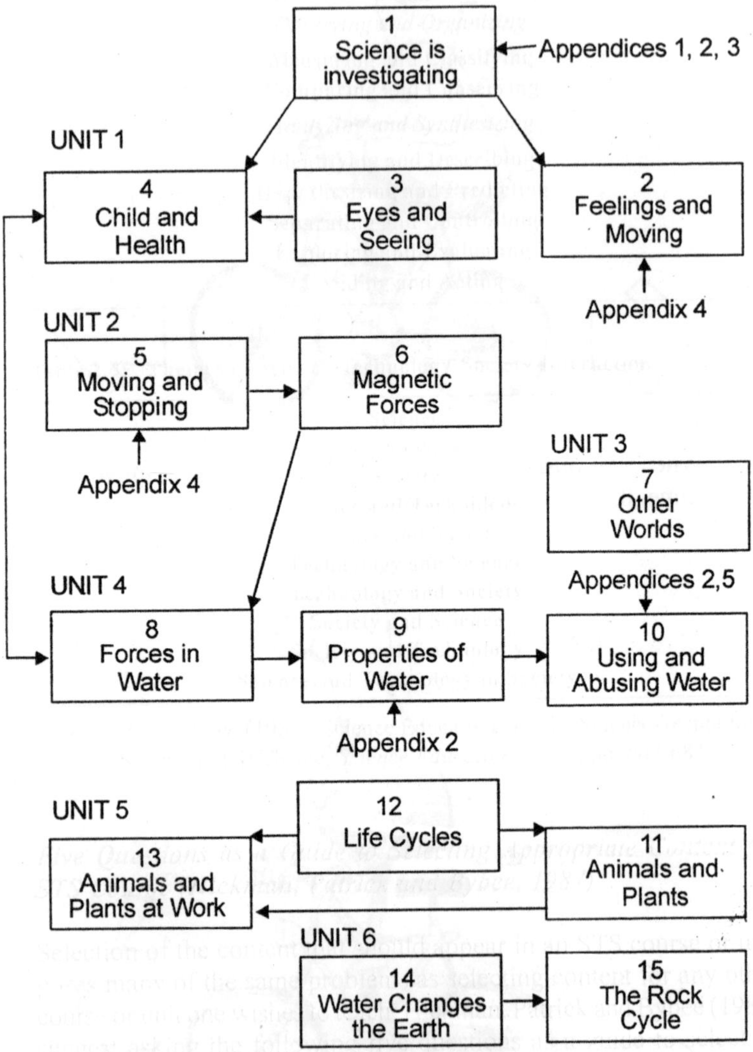

→ Concept or Process Links

Links between Chapters

Balancing Forces

When a force acts on an object, it moves. However, consider a tug-of-war. Here there are two equal forces in opposite directions[(a)]. There is no movement. Not until one force becomes bigger than the other does any movement occur[(b)].

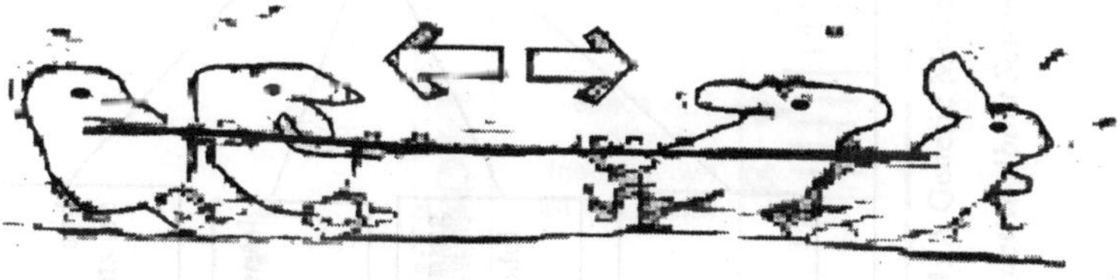

[(a)] Forces are balanced—no motion

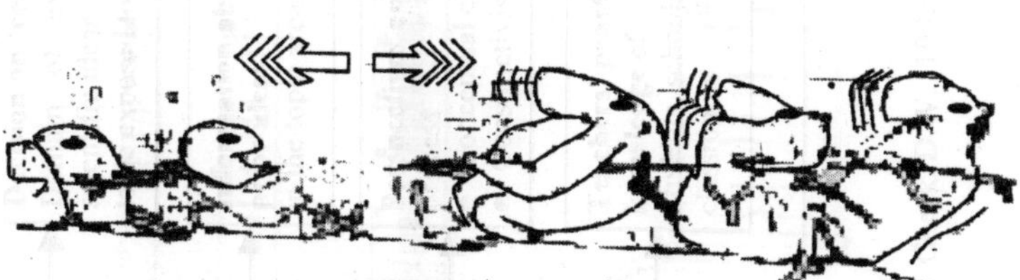

[(b)]Forces not balanced—motion

Balancing Forces

MEASURING LIQUID VOLUMES

Water and other liquids are measured in *litres*. This is usually written as 5 L, using a capital L for litre. The smaller unit of volume is the *millilitre*, written as mL.

$$1000 \text{ mL} = 1\text{L}$$

In the laboratory, you usually use only small volumes. Therefore you mostly use mL.

Volume is measured in *graduated cylinders* or *graduates*. Sometimes they are called *measuring cylinders*. These come in various sizes: 10 mL, 25 mL, 100 mL, 250 mL, 500 mL, and 100 mL (1L). You will mostly use the 10 mL, 25 mL, and 1000 mL sizes[1].

The 100 mL graduate is marked in 1 mL divisions. The space between two adjacent lines is 1 mL[2].

Suppose you want to measure 95 mL of water. Choose a 100 mL graduate. (A 10 mL graduate or a 100 mL graduate is not suitable.)

Pour some water into the graduate until the level is almost to the 95 mL mark. Notice the water surface is curved. This is the *meniscus*. It is this shape because the water sticks to the sides of the glass graduate.

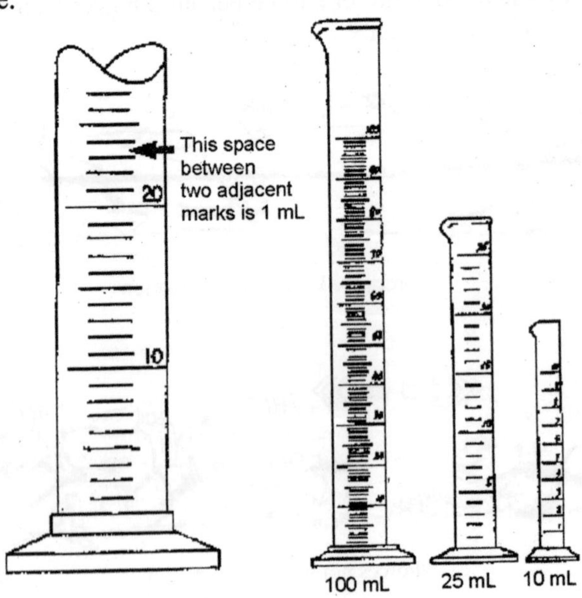

Chapter	1	2	3	4	5	6	7	8	9	10	11	12	13	14	15	Appendices
COMMUNICATING	←															→
OBSERVING	√				√	√				√	√	√		√	√	
INFERRING							√	√		√					√	
CLASSIFYING						√					√				√	
MEASURING	√		√		√											√
CALCULATING	√		√		√											
DRAWING GRAPHS		√			√											√
PREDICTING	√		√	√	√	√	√							√		
INTERPRETING DATA		√		√						√						√
USING WORKING DEFINITIONS										√						
CONTROLLING VARIABLES	√								√	√						
HYPOTHESISING (GENERALISING)	√		√	√	√	√			√					√		
EXPERIMENTING				√			√						√	√		
Using Models				√				√	√						√	
Using a Microscope										√	√	√				√
Using a Burner	√								√	√					√	√
Laboratory Skills	√								√	√			√		√	√

Grid of Processes and Skills

In the Lighter Vein

"Teacher, I know the present population figure. I wrote One billion because I thought you may correct this next week!"

APPENDIX H

Domain-wise Analysis of Lessons of Class VI, VII and VIII Science Textbooks

Table: Domain wise Analysis of the Chapters of CLASS VIth Science Text Book

	Chapter 1	Chapter 2	Chapter 3	Chapter 4	Chapter 5	Chapter 6	Chapter 7	Chapter 8	Chapter 9	Chapter 10	Chapter 11	Chapter 12	Chapter 13	Chapter 14	Total	Percentage
ACQUISITION OF KNOWLEDGE																
Concepts related to conventional content of Science																
Facts			0	2	6	1	21	0	16	0	0	6	1	0	53	5
Information	8	8	11	6	15	11	22	9	13	11	12	2	12	18	158	14.9
Concepts	8	0	12	9	15	10	0	8	0	6	4	0	5	14	91	8.6
Laws	0	0	0	0	0	0	0	0	0	0	0	0	0	0	0	28.5
Concepts related to Science and Technology																
Interaction & change			0	1	0	0	0	1	12	0	6	4	0	7	31	2.9
Growth & Cycles			0	2	0	0	0	5	14	4	5	2	1	2	35	3.3
Patterns & Processes	8	5	9	8	6	6	4	6	0	6	5	3	2	1	69	6.5
Probability & Prediction	1	2	0	1	0	0	0	0		0	0	1	0	0	5	0.47 13.17
Concepts related to STS																
Describe Interaction of ST&S			1	0	0	0	0	0	0	0	1	0	1	14	17	1.6
Understands limits of S&T			0	0	0	0	0	0	0	0	1	0	2	0	3	0.28
Distinguish Personal Opinion from Sc'ic Evidence			0	0	0	0	0	0	0	0	0	0	0	0	0	0
Understand Application of Technology			10	0	5	0	0	0	0	0	0	0	4	3	12	1.1 2.98

(Contd.)

(Contd.)

	Chapter 1	Chapter 2	Chapter 3	Chapter 4	Chapter 5	Chapter 6	Chapter 7	Chapter 8	Chapter 9	Chapter 10	Chapter 11	Chapter 12	Chapter 13	Chapter 14	Total	Percentage
SKILLS AND PROCESSES																
Questioning & Searching			10	5	0	5	5	2	2	6	5	5	4	6	65	6.1
Observing, using Space, time relp.			0	3	0	0	10	0	0	0	0	0	0	0	13	1.2
Classifying' Gr'ing & Org'ing	4	7	0	5	0	1	1		13	6	3	2	4	14	60	5.7
Interpreting Data	2	6	0	0	0	1	0	6	0	62	1	0	1	0	23	2.2
Comparing & Conserving	7	7	11	6	2	0	6	1	13	70	4	4	6	14	88	8.3
Critically Thinking Skills	1	0	0	0	0	0	0	9	0	0	2	0	0	0	12	1.1
Use Tech.in Practical Situation	1	0	11	0	6	0	0	0	0	0	2	0	7	3	30	2.8
Producing Mental Images	3	2	0	0	3	0	13	0	0	0	0	0	0	0	21	2.0
Combining objects & ideas in new ways		1	0	0	0	0	0	0	0	0	1	1	0	0	3	0.28
Predicting Possible consequenc-es		1	2	0	0	0	0	0	0		1	3	4	0	11	1
Suggesting possible causes of a given situation			0	0	0	0	0	0	0	3	1	2	5	0	11	1 31.7
ETHICS, VALUES AND ATTITUDES																
Understan-ding imp. of values and democratic process			0	0	0	0	0	0	0	3	0	0	0	0	3	0.28

(Contd.)

(Contd.)

	Chapter 1	Chapter 2	Chapter 3	Chapter 4	Chapter 5	Chapter 6	Chapter 7	Chapter 8	Chapter 9	Chapter 10	Chapter 11	Chapter 12	Chapter 13	Chapter 14	Total	Percentage
identifiying values bearing on any given STS issues like local issues, publc policies, global problems			0	0	0	0	0	0	0	0	1	0	3	4	8	0.76
Dev'ing positive attitude towards science in general, science in school & science teacher			0	0	0	0	0	0	0	0	0	0	0	0	0	0
Dev'ing positive attitude towards oneself			0	0	0	0	0	0	0	0	0	0	0	0	0	0
Expressing personal feeling in a constructive way			0	0	0	0	0	0	0	0	0	0	0	0	0	0
Developing sensitivity to and respect for the feeling of other people	0	0	0	0	0	0	0	0	0	0	0	0	0	0	0	0
Making decisions about social environment issues & exploring arguments on various sides of an issue			0	0	0	0	0	0	0	0	1	3	2	1	7	0.66 1.7

(Contd.)

(*Contd.*)

	Chapter 1	Chapter 2	Chapter 3	Chapter 4	Chapter 5	Chapter 6	Chapter 7	Chapter 8	Chapter 9	Chapter 10	Chapter 11	Chapter 12	Chapter 13	Chapter 14	Total	Percentage
APPLICATION AREAS																
Seeing instances of Scientific concepts in everyday life experiences	10	7	10	8	15	10	9	7	120	9	8	5	9	17	136	12.8
Undersatan-ding Science & T.principles as valued in household devices	1	1	5	0	0	9	0	0	0	0	0	0	0	12	28	2.6
Using scientific process in solving problems in everyday life	6		12	2	0	0	0	0	0	2	2	2	3	15	44	4.2
Making Decisions related to Personal health nutrition & life style based on knowledge of Scientific Concepts			2	1	0	0	0	0	0	9	1	3	1	1	18	1.7
Extending School experience beyond class room			0	0	0	0	0	0	0	0	0	0 –	0	0	0	0
Handling of apparatus			8	0	0	0	0	0	0	0	0	0		1	9	0.85
Development of Low cost material						0		0	0					0	0	22.25

Total = 1059

Table: Domainwise Analysis of the Chapters of CLASS VIIth Science Text Book

	Chapter 1	Chapter 2	Chapter 3	Chapter 4	Chapter 5	Chapter 6	Chapter 7	Chapter 8	Chapter 9	Chapter10	Chapter11	Chapter12	Chapter13	Chapter14	Chapter 15	Chapter16	Total	Percentage
ACQUISITION OF KNOWLEDGE **Concepts related to conventional content of Science**																		
Facts	0	1	0	0	1	1	0	1	0	1	0	0	0	4	0	0	9	1.3
Information	6	13	9	2	8	5	3	6	13	7	13	8	7	10	7	7	12 4	17.4
Concepts	3	13	2	3	1	0	1	2	3	0	0	2	0	6	0	0	36	5.1
Laws	2	0	0	0	1	0	0	2	2	0	0	0	0	0	0	0	7	0.98
																		24.78
Concepts related to Science and Technology																		
Interaction & change	0	2	0	0	0	0	0	0	3	0	4	1	0	0	5	0	15	2.1
Growth & Cycles	0	2	0	0	0	0	0	0	0	0	0	4	0	4	0	1	11	1.54
Patterns & Processes	0	12	1	2	1	4	2	5	7	3	5	10	7	5	2	2	68	9.5
Probability & Prediction	0	1	0	0	0	0	0	0	0	0	0	0	0	2	0	0	3	0.42
																		13.56
Concepts related to STS																		
Describe Interaction of ST&S	0	2	0	4	1	1	1	0	0	0	0	0	4	0	0	2	15	2.1
Understands limits of S&T	0	0	0	0	0	0	0	0	0	0	0	0	0	0	0	0	0	0
Distinguish Personal Opinion from Sc'ic Evidence	0	0	0	0	0	0	0	0	0	0	0	0	0	0	0	0	0	0
Understand Application of Technology	0	2	2	4	1	1	1	0	0	0	0	0	0	0	0	0	11	1.5
																		3.6
SKILLS AND PROCESS																		
Questioning & Searching	2	13	6	2	2	0	1	2	7	1	6	4	0	6	0	0	52	7.3
Observing,using Space,time relp.	0	7	0	0	0	0	0	0	0	0	0	0	0	0	0	0	7	0.98
Classifying Gr'ing & Org'ing	1	7	0	3	3	1	3	4	7	1	3	6	1	7	3	2	52	7.3
Interpreting Data	0	9	0	0	0	0	0	4	0	1	0	0	0	4	0	0	18	2.5
Comparing & Conserving	2	2	3	2	1	1	0	2	4	1	1	2	1	6	0	1	29	4.1

(Contd.)

(Contd.)

	Chapter 1	Chapter 2	Chapter 3	Chapter 4	Chapter 5	Chapter 6	Chapter 7	Chapter 8	Chapter 9	Chapter10	Chapter11	Chapter12	Chapter13	Chapter14	Chapter 15	Chapter16	Total	Percentage
Critically Thinking Skills	0	12	0	0	0	0	0	0	0	0	0	0	0	0	0	0	12	1.7
Use Tech. in Practical Situation	0	2	3	4	2	1	1	0	0	0	0	0	2	0	0	0	15	2.1
Producing Mental Images	0	0	0	0	0	0	0	0	0	0	0	0	0	0	0	0	0	0
Combining objects & ideas in new ways	0	0	0	0	0	0	0	0	0	0	0	0	0	0	0	0	0	0
Predicting Possible consequences	0	5	0	0	0	0	0	0	0	0	0	0	0	0	0	0	5	0.7
Suggesting possible causes of a given situation	0	4	0	0	0	0	0	1	0	0	0	0	0	0	0	0	5	0.7 27.38
ETHICS, VALUES AND ATTITUDES																		
Understanding imp. of values and democratic process	0	0	0	0	0	0	0	0	0	0	0	0	0	0	0	0	0	0
identifiying values bearing on any given STS issues like localissues,pub-lc policies,global problems	0	1	0	0	0	0	0	0	0	0	0	0	1	3	3	4	12	1.7
Dev'ing positive attitude towards science in general science in school & science teacher	0	0	0	0	0	0	0	0	0	0	0	0	0	0	0	0	0	0
Dev'ing positive attitude towards oneself	0	0	0	0	0	0	0	0	0	0	0	0	0	0	0	0	0	0
Expressing personal feeling in a constructive way	0	0	0	0	0	0	0	0	0	0	0	0	0	0	0	0	0	0
Developing sensitivity to and respect for the feeling of other people	0	0	0	0	0	0	0	0	0	0	0	0	0	0	0	0	0	0

(Contd.)

(Contd.)

	Chapter 1	Chapter 2	Chapter 3	Chapter 4	Chapter 5	Chapter 6	Chapter 7	Chapter 8	Chapter 9	Chapter10	Chapter11	Chapter12	Chapter13	Chapter14	Chapter 15	Chapter16	Total	Percentage
Making decisions about social environment issues & exploring arguments on various sides of an issue	0	0	0	0	0	0	0	0	0	0	0	0	1	1	6	1	9	1.3
																		3.0
APPLICATION AREAS																		
Seeing instances of Scientific concepts in everyday life experiences	8	13	6	5	5	2	3	1	2	1	4	7	7	5	5	0	74	10.4
Undersatanding Science & T.principles as valued in household devices	0	0	1	2	0	0	1	0	0	0	0	0	0	Q	0	0	4	0.56
Using scientific process in solving problems in everyday life	1	6	3	2	1	1	1	1	0	1	3	0	6	3	1	5	35	4.9
Making Decsions related to Personal health nutrition & life style based on knowledge of Scientific Concepts	0	5	1	4	0	0	0	0	0	0	0	1	6	5	4	7	33	4.6
Extending School experience beyond class room	0	4	0	0	0	0	0	0	0	0	0	0	0	4	0	0	8	1.1
Handling of apparatus	1	0	7	4	7	4	1	3	8	3	6	0	0	0	0	0	44	6.2
Development of Low cost material	0	0	0	0	0	0	0	0	0	0	0	0	0	0	0	0	0	0
																		27.76

Total = 713

Table: Domainwise Analysis of the Chapters of CLASS VIIIth Science Text Book

	Chapter 1	Chapter 2	Chapter 3	Chapter 4	Chapter 5	Chapter 6	Chapter 7	Chapter 8	Chapter 9	Chapter 10	Chapter 11	Chapter 12	Chapter 13	Chapter 14	Chapter 15	Chapter 16	Chapter 17	Chapter 18	Total	Percentage
ACQUISITION OF KNOWLEDGE Concepts related to conventional content of Science																				
Facts	4	0	0	0	1	1	0	0	0	0	1	0	0	0	2	1	3	0	13	1.5
Information	17	7	7	5	8	7	11	5	8	8	3	6	8	9	7	15	4	2	13 7	15.9
Concepts	15	4	3	1	5	2	2	1	1	3	1	0	4	4	3	0	0	0	49	5.7
Laws	5	2	0	0	0	0	0	0	0	0	0	0	4	0	0	0	0	0	11	1.3 24.4
Concepts related to Science and Technology																				
Interaction & change	0	0	0	0	4	1	1	4	1	0	0	3	0	1	1	0	0	0	16	1.9
Growth & Cycles	0	0	0	0	2	0	0	0	1	4	0	5	0	1	0	3	0	0	16	1.9
Patterns & Processes	14	6	6	4	6	5	2	0	0	5	3	2	3	5	1	2	1	0	65	7.5
Probability& Prediction	0	0	0	0	0	0	0	0	0	0	0	0	0	0	0	0	0	0	0	0 1 1.3
Concepts related to STS																				
Describe Interaction of ST&S	2	0	6	3	2	0	0	0	0	0	0	0	0	1	0	2	4	2	22	2.6
Understands limits of S&T	0	0	0	0	1	0	0	0	0	0	0	0	0	0	2	0	5	1	9	1.0
Distinguish Personal Opinion from Sc'ic Evidence	0	0	2	0	0	0	0	0	0	0	0	0	0	0	0	0	0	0	2	0.23
Understand Application of Technology	6	3	3	4	4	3	3	0	0	0	0	0	0	0	0	0	0	0	26	3.0 6.83
SKILLS AND PROCESSES																				
Questioning & Searching	4	5	3	1	0	0	2	4	3	0	0	0	0	0	0	1	0	1	24	2.8
Observing,us-ing Space,time relp.	0	0	0	0	0	0	0	0	0	0	0	0	0	0	0	0	3	0	3	0.35
Classifying, Gr'ing & Org'ing	0	0	0	0	3	1	5	4	4	4	2	0	4	10	8	6	0	1	52	6.0
Interpreting Data	4	6	3	0	2	0	0	0	0	0	0	0	1	0	0	0	0	0	16	1.9
Comparing & Conserving	1	2	2	0	6	2	3	2	4	2	2	2	4	1	3	1	0	1	38	4.4

(Contd.)

(Contd.)

	Chapter 1	Chapter 2	Chapter 3	Chapter 4	Chapter 5	Chapter 6	Chapter 7	Chapter 8	Chapter 9	Chapter 10	Chapter 11	Chapter 12	Chapter 13	Chapter 14	Chapter 15	Chapter 16	Chapter 17	Chapter 18	Total	Percentage
Critically Thinking Skills	5	0	0	0	0	0	0	0	1	0	0	0	0	0	0	0	0	0	6	0.7
Use Tech.in Practical Situation	0	4	6	3	4	3	0	2	4	0	0	0	2	6	8	2	2	2	48	5.6
Producing Mental Images	0	0	0	0	0	0	0	0	0	0	0	0	0	0	0	0	0	0	0	0
Combining objects & ideas in new ways	0	0	0	0	0	0	0	0	0	0	0	0	0	0	0	0	0	0	0	0
Predicting Possible consequences	0	0	0	0	0	0	0	0	0	0	0	0	0	0	0	0	0	0	0	0
Suggesting possible causes of a given situation	2	0	0	0	0	0	0	0	0	0	0	0	0	0	0	0	0	0	2	0.23
																				21.98
ETHICS, VALUES AND ATTITUDES																				
Understanding imp. of values and democratic process	0	0	0	0	0	0	0	0	0	0	0	0	0	0	0	0	0	0	0	0
Identifiying values bearing on any given STS issues like localissues, public policies,global problems	0	0	0	0	1	0	1	0	0	0	0	0	0	0	0	0	7	2	11	1.3
Dev'ing positive attitude towards science in general science in school & science teacher	0	0	0	0	0	0	0	0	0	0	0	0	0	0	0	0	0	1	1	0.12

(Contd.)

(*Contd.*)

	Chapter 1	Chapter 2	Chapter 3	Chapter 4	Chapter 5	Chapter 6	Chapter 7	Chapter 8	Chapter 9	Chapter 10	Chapter 11	Chapter 12	Chapter 13	Chapter 14	Chapter 15	Chapter 16	Chapter 17	Chapter 18	Total	Percentage
Dev'ing positive attitude towards oneself	0	0	0	0	0	0	0	0	0	0	0	0	0	0	0	0	0	0	0	0
Expressing personal feeling in a constructive way	0	0	0	0	0	0	0	0	0	0	0	0	0	0	0	0	0	0	0	0
Developing sensitivity to and respect for the feeling of other people	0	0	0	0	0	0	0	0	0	0	0	0	0	0	0	0	0	0	0	0
Making decisions about social environment issues & exploring arguments on various sides of an issue	0	0	0	0	0	0	0	0	2	0	0	0	0	0	7	2	7	2	20	2.3 3.72
APPLICATION AREAS																				
Seeing instances of Scientific concepts in everyday life experiences	13	5	5	3	4	6	12	6	4	3	3	6	10	2	1	15	3	2	103	11.6
Undersatanding Science & T. principles as valued in household devices	6	5	7	4	1	3	3	0	0	0	0	0	5	0	0	0	0	1	35	3.9
Using scientific process in solving problems in everyday life	10	4	6	2	1	4	6	3	1	1	1	0	2	0	4	2	1	2	50	5.62

(*Contd.*)

(*Contd.*)

	Chapter 1	Chapter 2	Chapter 3	Chapter 4	Chapter 5	Chapter 6	Chapter 7	Chapter 8	Chapter 9	Chapter 10	Chapter 11	Chapter 12	Chapter 13	Chapter 14	Chapter 15	Chapter 16	Chapter 17	Chapter 18	Total	Percentage
Making Decsions related to Personal health nutrition & life style based on knowledge of Scientific Concepts	1	0	1	0	1	2	5	2	7	2	0	0	1	2	7	13	4	2	50	5.62
Extend School experience beyond class room	10	0	0	0	0	0	0	1	0	0	0	0	0	0	0	0	0	0	11	1.24
Handling of apparatus	0	2	0	2	0	5	7	7	2	0	1	0	1	0	0	0	0	0	27	3
Development of Low cost material	0	0	0	0	0	0	0	0	0	0	0	0	0	0	0	0	0	0	0	0
																				31.9

Bibliography

Addinell, S. & Solomon, J. (1983). *Teacher's Guide—Science in a Social Context.* Hatfield, Herts., UK: Association for Science Education.

Ahmad, T. (1995). *A Study of Teachers' Awareness About Education for Science Technology Society and their Perception of Existing Science Content at Secondary Level in Achieving Identified ESTS Objectives*. Unpublished M. Phil. dissertation, Jamia Millia Islamia.

Aikenhead, G., Ryan, A. & Fleming, R. (1989). *Views on Science-Technology-Society*: Form CDN.mc5.Saskatoon, Saskatchewan, Canada, University of Saskatchewan, College of Education.

Aikenhead, G. (1975). *Science: A Way of Knowing.* University of Saskatchewan. Saskatoon.

Aikenhead, G. (1988). An Analysis of Four Ways of Assessing Students' Beliefs about STS Topics. *Journal of Research in Science Teaching*, 25, 607-629.

Aikenhead, G.S. (1985). Collective Decision Making in the Social Context of Science. *Science Education*, 69, 453-475.

Aikenhead, G.S. (1986). The Content of STS Education. *A Missive to the Science-Technology-Society Research Network*, 2(3), 18-23.

Aikenhead, G.S. (1987). High School Graduates' Beliefs about Science-Technology-Society. III. Characteristics and Limitations of Scientific Knowledge. *Science Education,* 71(4), 459-487.

Aikenhead, G.S. (1987). The Development of a New Technique for Monitoring Student Understanding of Science-Technology-Society Topics. *Curriculum Studies*, University of Saskatchewan, Canada.

Aikenhead, G.S. (1990). Scientific/Technological Literacy, Critical Reasoning, Classroom Practice. In L. Phillips & S. Norris (Eds.) *Foundations of Literacy Policy in Canada* (pp. 127-145). Calgary, Alberta: Dotselig Enterprises Ltd.

Aikenhead, G.S. (1992). The Integration of STS into Science Education. *Theory into Practice, 3(I),* 27-35.

Aikenhead, G.S. (1997) Toward a First Nations Cross-Cultural Science and Technology Curriculum. *Science Education*, 81(2), 219-237.

Aikenhead, G.S. and Ryan, A.G. (1992). The Development of a New Instrument: "Views on Science-Technology-Society" (VOSTS). *Science Education,* 76(5), 477-491.

Aikenhead, G.S. (1987). A Field Trial of a New Type of Assessment Instrument. *Project Report for Saskatchewan Schools*. College of Education, University of Saskatchewan.

Aikenhead, G.S. (1987). Views on Science-Technology-Society. Form CDN.mc.4. *Department of Curriculum Studies*, University of Sakatchewan, Canada.

Aikenhead, G.S., Fleming, R.W. & Ryan, A.G. (1987). High-School Graduates' Beliefs about Science-Technology-Society. Methods and Issues in Monitoring Students Views. *Science Education*, 71(2), 145-161.

Anderson, R.D. (1983). Are Yesterday's Goals Adequate for Tomorrow? *Science Education*, 67, 171-176.

Baez, A.V. (1986). *Science, Technology and Society: Educational Implications, in Innovations in Science and Technology Education*, Volume I, Layton D, (UNESCO), pp. 137-142.

Bell, B.F, Watts, D.M.; Ellington, K. (1985). *Learning Doing and Understanding in Science,* London; Secondary Science Curriculum Review.

Bhat, N.R. (1984). *To Formulate Objectives and Select Content for a Core Curriculum in Mathematics for Technicians*. Ph.D., Abstract in IV Survey of Research in Education 1983-88, Buch M.B., (ed.), NCERT, 1991.

Bingle, W. & Gaskell, J. (1994), Scientific Literacy for Decision-Making and the Social Construction of Scientific Knowledge. *Science education, 78(2),* 185-201.

Biological Science Curriculum Study. (1983). *Innovations*. Chicago: Kendall Hunt.

Blunck, S.M. & Yager, R.E. (1990). The Iowa chautauqua Program. A Model for Improving Science in Elementary School. *Journal of Elementary Science Education, 2,* (2), 3-9.

Boeker, E. (1979). *Science, Society and Education.* Vrije University, Amsterdam.

Bradford, C.S., Rubba, P.A. and Harkness, W.L. (1995). Views about Science-Technology-Society Interactions Held by College Students in General Education Physics and STS Course. *Science Education*, 79(4), 355-373

Brahadeeswaran, D. (1986). *An Analysis of the Effectiveness of Chemistry Curriculum of the Polytechnics*. Ph.D. Abstract in Fourth Survey of Research in Education 1983-88, Buch M.B., (ed.), NCERT, 1991.

Brinckerhoff, R.F. & Yager, R.E. (1986). *Exeter II Conference-Science and Technology Education for Tomorrow's World*. Washington, DC: National Science Teachers Association.

Bybee, R. (1991). Science-Technology-Society in Science Curriculum: The Policy-Practice Gap. *Theory into Practice, 30(4)*, 294-302.

Bybee, R.R. *et al.* (1980), Science, Society and Science Education, *Science Education*, Vol. 64 (3), 377-395.

Bybee, R. W. & Bonnstetter, R.J. (1987). What Research says: Implementing the Science-Technology-Society Theme in Science Education: Perceptions of Science Teachers. *School Science and Mathematics.* 87(2), 144-152.

Bybee, R.W. (1985). *Science-Technology-Society.* 1985 NSTA Year Book Washington DC, National Science Teachers' Association.

Bybee, R.W. (1986). The Sisyphean Question in Science Education: What Should the Scientifically and Technologically Literate Person Know, Value, and do as a Citizen? In *Science/Technology/Society.* Ed. Rodger W. Bybee. Washington, DC: National Science Teachers Association.

Bybee, R.W. (1987). Science Education and the Science-Technology-Society (S/T/S) Theme. *Science Education,* 71(5), 667-683.

Bybee, R.W. (1987). Teaching about Science-Technology-Society (STS): Views of Science Educators in the United States. *School Science and Mathematics,* 87(4), 274-285.

Bybee, R.W., Harms, N. Ward & Yager, R. (1980). Science, Society and Science Education. *Science Education,* 64(3), 337-395.

Bybee, R.W. and Mau, T. (1986). Science and Technology Related Global Problems: An International Survey of Science Educators. *Journal of Research in Science Teaching,* 23(7), 599-618.

Cambell, D.T. and Stanley, J.C.(1986*). Experimental and Quasi-Experimental Design for Research,* Chicago: Round McNaily and Co.

Case, R. (1985). *Intellectual Development: Birth to Adulthood.* New York: Academic Press.

Cheek, D. & Foltz, F. (1987). *A Comparison of Twelve Curricular Goal Statements for STS Education.* Unpublished paper. University Park, PA: STS Program, The Pennsylvania State University.

Cheek, D.W. (1992). *Thinking Constructively about Science, Technology and Society Education.* Albany, NY: State University of New York Press.

Curriculum of Middle Classes (1987). *Haryana Education Board,* Haryana.

Dave, P.N. and Dave, J. (1991). *Research in Curriculum—A Trend Report in IV Survey of Research of Education* (1983-1988), Buch M.B. (ed), NCERT.

Driver, R.H. & Oldham, V. (1986). A Constructive Approach to Curriculum Development in Science. *Studies in Science Education,* 13, 105-122.

Education Commission (1964-66). *Ministry of Education,* Government of India.

Edward, Z. (1991). Analysis of Student Scores in Fine Domains of Science Education for Concordance Across Domains and Growth within Domains Related to the Ability of Teachers to Pose Application Items During an S/T/S Intervention. *Dissertation Abstracts International* (1992), 53(1).

Eijkelhof, Harrie M.C. & Lijnse, P. (1988). The Role of Research and Development to Improve STS Education: Experiences from the PLON Project. *International Journal of Science Education,* 10(4), 464-474.

Eijkelhof, Harrie M.C. & Kortland, K. (1987). Physics in its Personal, Social and Scientific Context. *Bulletin of Science, Technology & Society,* 7(1/2), 126-136.

Eijkelhof, Harrie M.C. & Verhagen, P.A.J. (1985). A Thematic Approach to Curriculum Development in Senior High Schools. In *The Many Faces of Teaching and Learning Mechanics.* Ed. P.L. Lijnse. Utrecht: WCC Publishers.

Eikelhof, H.M.C. *et al.* (1982*)* Physics in Society, *Bulletin of Science, Technology and Society* Vol. 2(1), 59-101.

Erickson, G.L. (1979). Children's Conceptions of Heat and Temperature. *Science Education 64*(3), 323-336.

Fensham, P.J. (1976) "Social Content in Chemistry Courses", *Chemistry in Britain* 12(5), 148-151.

Fleming, R.W. (1986). Adolescent Reasoning in Socio-Scientific Issues. I. Social Cognition. II. Non-Social Cognition. *Journal of Research Science Teaching*, 23(8), 677-698.

Fleming, R.W. (1987). High School Graduate Belief about Science-Technology-Society. The Interaction among Science, Technology and Society. *Science Education.* 71(2), 163-186.

Fleming, R. (1988). Undergraduate Science Students' Views on the Relationship between Science-Technology-Society. *International Journal of Science Education, 10, 449-463.*

Fleming, Reginald W. (1987). *Technological Literacy.* Regina, Canada: Department of Education.

Fullick, P. (1992). Addressing Science and Technology Issues in the United Kingdom: The SATIS Project. *Theory into Practice,* Vol. XXXI(1), 36-43.

Good, R., Kromhout, R., Lawson, A., and Renner, J. (1985) *Science education: Definitions and Implications for Research.* Paper presented at the annual meeting of the National Association for Research in Science Teaching. French Lick, Indiana.

Goodlad, J.I. (1984). *A Place Called School: Prospects for the Future.* New York: McGraw Hill.

Hall, W. *et al.* (1983). *Teaching Science, Technology and Society in the Junior High School,* Brisbane, Queensland, Australia.

Hansen, K.H. and Olsen, J. (1996). How Teachers Consture Curriculum Integration: The Science, Technology, Society (STS) Movements as Bildung. *Curriculum Studies*, 28(6), 669-682.

Harms, N.C. (1981). Project Synthesis: Summary and Implications for Teachers. In *What Research Says to the Science Teacher.* Vol. 3.Eds. Norris C. Harms, & Robert E. Yager. Washington, DC: National Science Teachers Association.

Harms, N.C. (1977). *Project Synthesis: An Interpretative Consolidation of Research Identifying Needs in Natural Science Education.* Proposal prepared for the National Science Foundation. Boulder Co., University of Colorado.

Harms, N.C. and R.E. Yager (1981). *What Research Says to the Science Teacher* (Vol. 3). Washington, DC: National Science Teachers Association.

Hart, E.P. & Robottom, I.M. (1990). The Science-Technology-Society Movement in Science Education: A Critique of the Reform Process. *Journal of Research in Science Teaching, 27*, 575-588.

Heath, P. (1992). Organizing for STS Teaching and Learning: The Doing of STS. *Theory into Practice, 31(I),* 52-58.

Helgeson, S.L., Blosser, P.E. and Howe. R.W. (1977). *The Status of Pre-College Science, Mathematics, and Social Science Education: 1955-75.* Columbus, Ohio: Centre for Science and Mathematics Education, The Ohio State University.

Hickman, F. Patrick, J.J., & Bybee, R.W. (1987). *Science/Technology/Society: A Framework for Curriculum Reform in Secondary Science and Social Studies.* Boulder, CO: Social Science Education Consortium Inc.

Hofstein, A, Yager, R.E (1982). Societal Issues as Organizers for Science Education in the 80s. *S School Science and Mathematics.* 82(7), 539-547.

Holman, J.S. (1985b). Science and Technology in society. A new initiative. *Education in Science,* 111-44.

Hoshangabad Vigyan (1977). Science Today, December 1977, 13-33.

Hungerford, H.R., Peyton, R.B., & Wilkey, R.J. (1980). Goals for curriculum development in environment education. *Journal of Environmental Education,* 11(3), 42-47.

Hurd, P. (1985). A Rationale for a Science, Technology, and Society Theme in Science Education. In R.W. Bybee (Ed.). *Science, Technology, Society 1985 Year-book of the National Science Teachers Association* (pp. 94-101). Washington DC.

Hurd, P.D. (1984). *Reforming Science Education:* The Search for a New Vision. Washington, DC: Council for Basic Education.

Hurd, P.D. (1975) Science, Technology and Society; New Goals for Interdisciplinary Teaching, *The Science Teacher*, Vol. 42: p. 27-30.

Hurd, P.D. (1981). Biology Education. In N.C. Harms & R.E. Yager (Eds.), *What research says to the science teacher* (Vol. 3, pp. 12-32). Washington, DC: National Science Teachers Association.

Iozzi, L.A. (1982). *Curriculum Model for Preparing for Tomorrow's World.* Longmont, CO: Sopris West Inc.

Iozzi, L.A. (1987), Science-Technology-Society: *Preparing for Tomorrow's World. Teacher's Guide*. Longmont, CO: Sopris West Inc.

Iskandar, S.M. (1991). An Evaluation of the Science-Technology-Society Approach to Science Teaching. *Dissertation Abstracts International* (1992), 53(1).

Jager, H. & Vander Loo, F. (1990a). Decision-Making in Environmental Education-Notes from Research in the Dutch NME-VO Project. In *Technological Literacy V: Proceedings of the Fifth National Technological Literacy Conference,* February 2-4, 1990, Arlington, VA. Ed. Dennis W. Cheek. Bloomington, IN: ERIC Clearing House for Social Studies/Social Science Education.

Jalota. S.S. (1976) *Mansik Yogyta Ki Sansodhit Samuhik Pariksha, Manikaran Tatha Pariksha Ki Vidhi (Manual)*, Chandigarh.

Jarcho, I.S. (1986). Science, Technology and Society Education: Methods and Materials. *Bulletin of Science, Technology and Society*, 6(2/3), 282-286.

Jarvis, T. and Rennie, L.J. (1996). Understanding Technology: the Development of a concept. *International Journal of Science Education*, 18(8), 977-992.

Johnson, D.W. & Johnson, R. (1975). *Learning Together and About: Cooperation, Competition and Individualization*. Englewood Cliffs, N.J.: Prentic, Hall.

Kakshawar Pathyakaram Ke Masik Baat (1993). Secondary Siksha Vibhag, Chandigarh, Government of Haryana.

Karplus, R. (1977). *Science Teaching and the Development of Reasoning*. Berkeley, CA: University of California Press.

Khan, J.M. (1987). *Attitudes of Policy Makers Towards Science And Society Issues In A Developing Country* Paper Presented in 4th International IOSTE Symposium on world Trends In Science and Technology Education held at Kiel, Germany, 4-12 August, 1987.

Klopfer, Shoresman, Carrier & Gies (1970) *Test on Understanding Science Form Ev*. Learning, Research and Development Centre, University of Pittsburg.

Kortland, K. (1996). The STS Case Study About Students. Decision Making on the Waste Issue. *Science Education*, 80(6), 673, 689.

Layton, D. (1986). Empowerment of People: The Educational Challenge of Science for Specific Social Purposes (SSSP). *Bulletin of Science, Technology & Society*, 6(2/3), 210-218.

Layton, D. (1988). Revaluing the T in STS. *International Journal of Science Education*, 10(4), 367-79.

Levin, F.S. & Lindbeck, J.S. (1979). An Analysis of Selected Biology Textbooks for the Treatment of Controversial Issues and Biosocial Problems. *Journal of Research in Science Teaching*, 16(3), 199-203.

Lewis, J. (1978), Science in Society, *Physics Education*, Vol. 13(6): 340-343, 1978.

Lewis, J.L. (ed) (1987). *Science and Technology Education and Future Human Needs, Vol.1 of Science and Technology and Future Human Needs*. Oxford: Pergamon Press for International Council of Scientific Unions Press.

Lowe, I. (1982). 'Establishing STS at High School and University' Paper presented at New Zealand symposium: '*Education About Science: Critical Challenges to Science Education*, Wellington, New Zealand, September.

Lu, Yu-Ling (1994). A Study of the Effectiveness of Science-Technology-Society Approach to Science Teaching in the Elementary School. *Dissertation Abstracts International*, 54(9), 3390-A.

MacCleave-Frazier, A. & Murray, E.C. (1984). A Framework for Reconceptualizing Home Economics. *Canadian Home Economics Journal*, 34(2), 69-73.

Masih, A. (1998). *New Trends in Science Curriculum*, New Delhi, Manak Publishing Pvt. Ltd.

McComas, W.F. (1989). Changing Student Attitudes Through S/T/S Education: The Results of the 1987-88 Chautauqua Workshops. *Chautauqua Notes*, 5(1), 1-3.

McComas, W.F. (1989). Just the Facts: The Result of the 1987-88 Chautauqua Workshops, *Chautauqua Notes, 4,* (4), 1-2.

McComas, W.F. and Yager, R.E. (1988). *The Iowa Assessment Package for Evaluation in Five Domains of Science Education.* Iowa City, IA, University of Iowa, Science Education Center.

McConnell, M. (1982). Teaching About Science Technology and Society at the Secondary School Level in the United States. An educational dilemma for the 1980s. *Studies in Science Education,* (9), 1-32.

McFadden, C.P. (1991). Towards an STS School Curriculum. *Science Education,* 75(4), 457-469.

Mckenzie, K.V. (1983). *Science Technology Society: An Maganizing Framework for Content Selection Teaching Science Technology and Society in the High School,* BCAE, Griffith University.

Meichtery, Y.J. (1993). The Impact of Science Curricula on Student Views About the Notion of Science. *Journal of Research in Science Teaching,* 30(5), 429-443.

Menon, S.B. (1986). *The Study of a System of Science Education in the Perspective of the Process of Science Inquiry,* Ph.D. abstract in IV Survey of Research in Education, 1983-1988, Buch, M.B. (ed) pp.738-739, NCERT, 1991.

Meyer, J. (1988). Themes of Social Responsibility: A Survey of Three Professional Schools. *Journal of Business Ethics,* 7, 313-320.

Meyer, R. (1977). 'Science Curricula in Australia-A National View', *Australian Science Teacher Journal 23 (3)* November, pp. 90-97.

Mitchener, C.P. and Anderson, R.D. (1989). Teachers' Perspective: Developing and Implementing and Implementing an STS Curriculum. *Journal of Research in Science Teaching,* 26(4), 351-369.

Moore, R.W. and Sutman, F. (1970). The Development, Field Test And Validation of An Inventory of Scientific Attitudes. *Journal of Research In Science Teaching,* 7(1) 85-94.

Morland, M. *et al.* (1981), North West Minster Community *School: Science, Technology and Society* (STAS) London, U.K.

Myers, L.H. (1989). *Analysis of Students' Outcomes in Ninth Grade Physical Science Taught with a Science-Technology-Society focus verses one Taught with a Textbook Orientation.* Doctoral Dissertation. The University of Iowa.

National Assessment of Education Progress (1978). *The third assessment of science,* 1976-77 (Report No. 08-5-08). Denver, CO: Author.

National Association for Research in Science Teaching (1985). A Symposium on *"Science Education: Definitions & Implications for Research"*, French Lick Springs, Indiana, April.

National Council for the Social Studies (1990a). Teaching About Science, Technology and Society in Social Studies: Education for citizenship in the 21st century. *Social Education,* 54(4), 189-193.

National Curriculum for Elementary & Secondary Education: A Framework (1988), NCERT, New Delhi.

National Curriculum for Primary and Secondary Education—A Framework (1985). NCERT, New Delhi.

National Curriculum Framework for School Education—A Discussion Document (2000), NCERT, New Delhi.

National Policy On Education-1986. Ministry of Human Resource Development. Department of Education. Government of India.

National Science Teachers Association, (1990). *The NSTA Position Statement on Science/Technology/Society (STS).* Washington, DC: NSTA.

National Science Teachers Association, (1982). *Science-Technology-Society: Science Education for the 1980s.* Washington, DC: National Science Teachers Association.

New York State Education Department, (1985). *Science, Technology and Society-Block J: Science Syllabus for Middle and Junior High Schools.* Albany, NY: author.

Northern Examination and Assessment Board (1995). *Syllabus for Science, Technology and Society* (Advancement Supplementary), U.K.

Northern Examining Association (1988). *Science, Technology and Society-Syllabus for the 1988 Examination.* General Certificate of Secondary Education, Syllabus for Mature Students. Hull, England: Northern Examining Association.

Nussbaum, J. & Novick, S. (1981b). Brainstorming in the Classroom to Invent a Model: A Case Study *School Science Review,* 62(221), 771-778.

Okebukola, P.A. & Jegede, O.J. (1990). Eco-cultural Influences upon Students' Concept Attainment in Science, *Journal of Research in Science Teaching,* 27(7), 661-669.

Osborne, Roger & Wittrock, M.C. (1983). Learning Science: A Generative Process. *Science Education,* 67(4), 489-508.

Parashar, M. (1994). *STS Approach in Teaching of Science: Implications or Science Education in India.* Unpublished M.Phil. Dissertation, Department of Education, University of Delhi.

Pedretti, E. & Hodson, D. (1995). From Rhetoric to Action: Implementing STS Education through Action Research. *Journal of Research in Science Teaching,* 32(5) 463-486.

Pedretti, E. (1996). Learning about Science, Technology and Society (STS) through an Action Research Project: Co-constructing an Issue-based Model for STS Education. *School Science and Mathematics,* 96(8), 432-40.

Pedretti, E. (1997). Septic Tank Crisis: A Case Study of Science, Technology and Society Education in an Elementary school. *International Journal of Science Education,* 12(10), 1211-1230.

Pedretti, E.G. (1997). Action Research in Science-Technology-Society Education: The Road Less Traveled. *Dissertation Abstracts International* 55(12), 3801-A

Penick, J.E. and Yager R.E. (1986). Trends in Science Education. *European Journal of Science Education,* 8(1), p. 1-8.

Piel, E.J. (1981). Interaction of science, technology, and society in secondary School. In N.C. Harms & R.E. Yager (Eds.), *What research says to the science teacher* (Vol. 3, pp. 94-112). Washington, DC: National Science Teachers Association.

Piel, J. (1990). *Science/Technology/Society interactions: Teaching in the secondary school.* In New Trends in Integrated Science Teaching, UNESCO, Paris.

PLON (1972). Project at the University Utrecht. Nether land.

Price, R.F. & Cross, R.T. (1995). Conceptions of Science and Technology Clarified: Improving the Teaching of Science. *International Journal of Science Education 17(3)*, 285-293.

Program of Action (1992). MHRD, New Delhi.

Programme of Action, National Policy on Education, 1986. Ministry of Human Resource Development. Department of Education. Government of India.

Ramanathan, S. (1993). *An Assessment of The Representation of Science in NCERT Upper Primary Science Textbooks.* M.Phil. Dissertation. University of Delhi.

Ramsey, J. (1994). The Science Education Reform Movement: Implications for Social Responsibility. *Science Education*, 77(2), 235-258.

Renner, John W. (1982). The Power of Purpose. *Science Education,* 66(5), 709-716.

Rosenthal, D.B. (1984). Social Issues in High School Biology Textbooks: 1963-1983. *Journal of Research in Science Teaching,* 21(8), 819-831.

Roy, S. (1988). *A Critical Evaluation of the High School General Science Textbooks in Bangladesh,* Ph.D. Abstract 1988, *in* IV Survey of Research in Education 1983-1988, Buch M.B., (ed), NCERT, 1991.

Roy, R. (1983). Math and Science Education: Goal not included. *The Christian Science Monitor,* pp. 36-37.

Roy, R. (1985a). The Science/Technology/Society Connection. *Curriculum Review.* 24(3), 13-1.

Roy, R. (1985b). Bright Vision/Impossible Dream. *S-STS Reporter*, 1-2.

Rubba, A.P. and William, L.H. (1993) Examination of Pre-service and In-service Secondary Science Teachers Beliefs about Science-Technology-Society Interactions. *Science Education*, 77(4), 407-431.

Rubba, P. (1991). Integrating STS into School Science and Teacher Education: Beyond Awareness. *Theory into Practice, 30(4),* 303-308.

Rubba, P.A. (1976). *Nature of Scientific Knowledge Scale.* Unpublished Manuscript, Bloomington: Indiana University, School of Education.

Rubba, P.A. & Wiesenmayer, R.L. (1988). Goals and Competencies for Pre-college STS Education: Recommendations Based upon Recent Literature in Environmental Education. *Journal of Environmental Education*, 19(4), 38-44.

Rubba, P.A. (1987a). The Current State of Research in Pre-college STS Education: A Position Paper, *Bulletin of Science, Technology & Society,* 7(1/2), 248-252.

Rubba, P.A. (1989). An Investigation of the Semantic Meaning Assigned to Concepts Affiliated with STS Education and of STS Instructional Practices Among a Sample of Exemplary Science Teachers. *Journal of Research in Science Teaching*, 26(8), 687-702.

Rubba, P.A. & Harkness. W.L. (1993). Examination of Pre-service and In-service Secondary Science Teachers' Beliefs about Science-Technology-Society Interactions. *Science Education*, 8, 1-8.

Rubba, P.A., Bradford, C.S. & Harkness, W. J.(1996). A New Scoring Procedure for the Views on Science-Technology-Society Instrument. *International Journal of Science Education,* 18(4), 387-400.

Rubba,. P.A. & Wiesenmayer, R.L. (1985). A Goal Structure for Pre-college STS Education: A Proposal Based upon Recent Literature in Environmental Education. *Bulletin of Science, Technology & Society,* 5(6), 573-580.

Rubba, P.A.; McGuyer, M. and Wahlund, T.M. (1991). The Effects of Infusing STS Vignettes into the Genetics unit of Biology on Learner Outcomes in STS and Genetics: A Report of Two Investigators. *Journal of Research in Science Teaching,* 28(6), 537-552.

Ryan, A.G. (1987). High School Graduates' Beliefs about STS. The Characteristics of Scientists, *Science Education,* 71, 489-510.

Ryan, A.G. & Aikenhead, G.S. (1992). Students' Preconceptions about the Epistemology of Science. Science Education, 76(6), 559-580.

Sarangapani, P. (1990). *A Critique of Primary School Mathematics Curriculum from a Piagetian Perspective,* Unpublished M.Phil Thesis, Delhi University.

SATIS, (1991). *The Framework Units* (Hunt. A. and Solomon. J). Association for Science Education. Hatfield, 16-19.

Science and Society Committee, (1990). Teaching about Science, Technology and Society in Social Studies, Education for Citizenship in the 21st Century. *Social Education,* 54, 189-193.

Science through Science, Technology and Society (S-STS) Project, (1985). Task force establishes materials review process. *S-STS Reporter*, 1(3), 5-6.

SISCON, (1983) Science in a social context, Solomon, J. Hatfield, UK Association for Science Education. UNESCO (1990). New Trends in Integrated Science Teaching Vol. VI, UNESCO, Paris.

Sixth All India Educational Survey, (1998), Selected statistics, National Council of Educational Research and Training.

Skilbeck, M. (1984). *School-Based Curriculum Development*. Harper. London.

Solbes, Joan & Vilches, A (1997), STS Interactions and the Teaching of Physics and Chemistry. 16-19, 81(2), 377-386.

Solomon, J. (1988b). The Dilemma of Science, Technology, and Society Education. In *Development and Dilemmas in Science Education.* Ed. Peter J. Fensham. Philadelphia, PA: The Falmer Press.

Solomon, J. (1980b). Science and Society Studies in the School Curriculum. *Social Science Review,* 62, 213-219.

Solomon, J. (1993). *Teaching science, technology and society.* Philadelphia: Open University Press.

Soong, B.C. & Yager, R.E. (1993). The Inclusion of STS Material in the most Frequently used Secondary Science Textbooks in the US. *Journal of Research in Science Teaching,* 30(4), 339-349.

Souque, J.P. (1987). Science Education and Textbook Science. *Canadian Journal of Education,* 12(1), 74-86.

Stake, R.E. & Easely, J. (1978). *Case Studies in science education.* (Vols. I and II). Washington, DC:U.S. Government Printing office.

Syllabus for upper primary stage (1987). NCERT, New Delhi.

Taba, H. (1962). Curriculum Development: *Theory and Practice*. New York: Harcourt, Brace and World Inc.

Tamir, P. (1987). *Science and Society in the Second IEA Science Study: An Israeli Perspective.* In K. Riquarts (ed.), Science and Technology Education and Quality of Life, Vol. 2, pp. 626-639.

Technology Policy Statement (1983). Department of Science and Technology, Government of India, New Delhi.

Thier, H. & Nagle, B. (1994). Developing a Model for Issue Oriented Science. In J. Solomon & G. Aikenhead (Eds.), *STS Education: International Perspectives on Reform*. (pp. 75-83). New York: Teachers' College Press.

Thirunarayanan, M.D. (1990). *Students' Conceptions of Selected Science, Technology and Society Issues: An Exploratory Multi-dimensional Scaling Study.* Ph.D. Thesis, Arizona State University.

Ventura, A. (1982). *A Report on the Status of Science and Technology in Botswana and Recommendations for their Improvement.* UN Financing System for Science and Technology for Development, New York.

Vigyan, (1994). *Text Book for Class VIII*, Haryana Education Board, Haryana.

Vigyan, (1994). *Text Book for Class VI*, Haryana Education Board, Haryana.

Vigyan, (1994). *Text Book for Class VII,* Haryana Education Board, Haryana.

Waks, L.J. (1986). Reflections on Technological Literacy. *Bulletin of Science, Technology & Society,* 7(1/2), 357-366.

Waks, L.J. (1992). The Responsibility Spiral: A Curriculum Framework for STS Education. *Theory into Practice*, 31, 13-19.

Waks, L.J. & Barchi, B.A. (1992). STS in U.S. School Science: Perception of Selected Leaders and Their Implications for STS Education. *Science Education*, 76(1), 79-90.

Waks, Leonard J. (1987a). A Technological Literacy Credo. *Bulletin of Science, Technology & Society,* 7(1/2), 357-366.

Weiss, I.R. (1978). *Report of the 1977 National Survey of Science, Mathematics, and Social Studies Education: Center for Educational Research and Evaluation.* Washington, DC: United States Government Printing Office.

Welch, W.W. (1966). *Welch Science Process Inventory*, Form D. Minneapolis: Minnesota Research and Evaluation Center, University of Minnesota.

World Trends in Science and Technology Education (1987). Ed. Kurt Riquarts, Kiel, Germany.

Yager, R.E. (1993). Science-Technology-Society as Reform. School Science & Mathematics, 145-151.

Yager, R. (1992). Viewpoint: What We did not Team from the 60s about Science Curriculum Reform ? *Journal of Research in Science Teaching, 29(8)*, 905-910.

Yager, R. (1980). *Analysis of Current Accomplishments and Needs in Science Education.* ERIC/SMEAC Clearinghouse for Science, Mathematics and Environmental Education. Ohio State University, Columbus.

Yager, R. (1984). Defining the Discipline of Science Education. *Science Education*, 68(1), 35-37.

Yager, R. (1985). An Alternate View. *Journal of College Science Teaching*, 14(3), 223-224.

Yager, R. (1985). In Defense of Defining Science Education as the Science/ Society Interface. *Science Education,* 69(2), 143-144.

Yager, R.E. & Broenkhorst, B.J. (1987). The IOWA STS Project: Student Growth in Variety of Domains. Paper Presented in the 4th International IOSTE Symposium on World Trends in Science and Technology Education, Kiel, Germany.

Yager, R.E. (1984). The Major Crisis in Science Education. *School Science and Mathematics,* 83, 189-198.

Yager, R.E. (1985b). STS–Something New in Education. *Bulletin of Science, Technology & Society,* 5(6), 568-572.

Yager, R.E., Blunck, S.M. & Ajam, M. (1990). *The Iowa Assessment Package for Evaluation in Five Domains of Science Education.* The University of Iowa.

Yager, R.E., Hofstein, A. & Lunetta, V.N. (1981). Science Education Attuned to Social Issues: Challenges for the 80s. *The Science Teacher,* 48(9), 12-14.

Yager, R.E. & Hofstein, Avi (1986). Features of a Quality Classroom for School Science. *Journal of Curriculum Studies,* 18(2), 115-132.

Yager, R.E. & McCormack, A.J. (1989). Assessing Teaching/Learning Successes in Multiple Domains of Science and Science Education. *Science Education* 73(1): 45-58.

Yager, R.E. (1989). Development of Student Creative Skills: A Quest for Successful Science Education. *Creativity Research Journal 2(3):* 196-203.

Yager, R.E. & Penick, J.E. (1988). Change in Perceived Attitudes Towards the Goals for Science Instruction in Schools. *Journal of Research in Science Teaching, 25, (3), 179-184.*

Yager, R.E. & Tamir, P. (1993). STS Approach: Reasons, intentions, accomplishments, and outcomes. *Science Education*, 77(6), pp. 637-658.

Yager, R.E. (1990). *Science, Technology & Society (STS): Trends in the Integration of Science Education.* In New Trends in Integrated Science Teaching, UNESCO, Paris.

Yager, R.E., & J.E. Penick, (1984). What Students Say about Science Teaching and Science Teachers. *Science Education 68(2):* 143-152.

Yager, R.E., & J.E. Penick, (1986). Perceptions of Four Age Groups Toward Science Classes, Teachers, and the Value of Science, *Science Education* 70(4): 355-363.

Yager, R.E., Blunck, S.B. and Ajam. M. (1990). *The Iowa Assessment Package for Evaluation in Five Domains of Science Education.* (2nd. Ed.). Iowa City, IA: The University of Iowa, Science Education Center.

Yash Pal Committee, (1993). *Learning Without Burden – Report of the National Advisory Committee.* MHRD, New Delhi.

Yoshisuke, K. (1993). The Effects of STS Instruction in Japan Compared to Results Reported in the US. *Dissertation Abstracts International* (1994), 55(3), 524-A.

Young, M. (1971). An Approach to the Study of Curricula as Socially Organized Knowledge. In M. Young (ed). *Knowledge and Control.* Collier-Macmillan, London.

Young, M. (1988). The Selection of Processes, Context and Concepts and their Relation to Methods of Teaching, in Harlen W. (ed), *New Trends in Primary School Science Education,* pp. 7-16, Paris, UNESCO.

Zachariah, M. & Sooryamoorthy, R. (1994). *Science for Social Revolution. The Kerela Sastra Sahitya Parishad.* Vistaar Publications, New Delhi.

Ziman, J. (1980). *Teaching and Learning about Science Society.* Cambridge: Cambridge University Press.

Zoller *et al.* (1990). Goal Attainment in Science Technology-Society (S/T/S) Education and Reality: The Case of British Columbia. *Science Education,* 74(1), 19-36.

Zoller, U. (1987). Problem-solving and Decision-making in Science-Technology-Environment-Society (STES) Education. In K. Riquarts (ed.), *Science and Technology Education and the Quality of Life,* Vol. 2 (Kiel: IPN), 562-69.

Zoller, U. (1991). Teaching/Learning Styles, Performances, and Students' Teaching Evaluation in S/S/E/S-Focussed Science Teacher Education: A Quasi-quantitative Probe of a Case Study. *Journal of Research in Science Teaching,* 28(7), 593-607.

Index